I07539447

The Hall of Records.

A MUNICIPAL EXPERIMENT

or

THE HALL OF RECORDS POWER PLANT

by

REGINALD PELHAM BOLTON

—

1917

—

NEW YORK

THE BUREAU OF PUBLIC SERVICE ECONOMICS, INC.

55 LIBERTY STREET

Reader,

Some theories and policies are in much the same situation as a baseball—it has been hit so hard by both sides interested that it is most of the time out of sight of the disinterested observer.

Hopeful anticipations and pleasing illusions must be brought to earth to enable ordinary people to discover their value for practical assimilation.

The worth of a Policy, such as that of Municipal Ownership, or a Method, such as that of the operation of isolated plants, is not to be determined by mere argument or assertion, but by proven value and positive results. Thoughtful minds ought not to be expected to accept any principle as sound that is incapable of supporting itself by practical achievements.

The acid test of any economic problem is its financial outcome, and, applying to such matters the age-old advice of a wise student of human affairs, the best method of dealing with such problems is to

Prove all things.

The Author.

ILLUSTRATIONS

CHAPTER I.

THE TEST AND ITS LESSONS.

DURING the year 1913, a remarkable trial was conducted in the important building known as the Hall of Records, in the City of New York, to determine the cost, including all financial elements, of installing and operating the mechanical power plant which is in that building, by which the lighting, the heating, and the elevator service of the Hall of Records, and of several nearby buildings had been for some time provided.

The result of this trial, which was unusually extensive, and included most interesting details, has not heretofore been published, although an exceedingly voluminous report has been issued by one of the parties engaged in the trial, which includes practically all the documents, papers, computations and records which were produced during the trial, and the correspondence and discussions of the several parties interested in the operation.

This report, however, is so excessively voluminous that in itself it precludes a clear understanding by the ordinary reader of the essential features of the trial, and the definite results established thereby; nor does it include a discussion of some of the most interesting economical, financial and physical results derived from the trial.

These are now to be presented in a more compact, and perhaps more readily understandable form.

The particular interest attaching to the test is the establishment of the economic conditions under which a first-class isolated power-plant may be operated, when its control is placed in the hands of thoroughly trained engineers, and its operation is directed by them with a special effort towards the highest economy, and the least possible current expense. Moreover, this plant, although designed and installed for the service only of the building in which it is placed, was afforded the opportunity during the trial of increasing its output so as to provide light and heat to other buildings in the vicinity, thus largely increasing its opportunity for economic results, since the plant was of ample capacity for such extended service.

The trial also determined the capital expense to which a municipality is put in the establishment of such plants and the fixed charges that must be provided by the taxpayers upon money borrowed for its purchase. Further, the comparative economy of the purchase by a municipality of service from public systems of steam and electric supplies, was disclosed in a conclusive manner. The facts for these determinations have heretofore been limited to observations over a short period of time. In this trial they included every physical and financial element existing during a whole year.

This test has, therefore, a very unusual value, in its presentation of all the features which enter into a municipal ownership operation of a character which can be readily understood, and which involves less complexities than larger projects, such as the operation of public utilities and railways.

The circumstances which led to the trial were somewhat involved, and the decision to conduct the test was the outcome of discussions, which had extended over a considerable period of time, between the city officials responsible for the operation of this and other municipal buildings, the particular department of the city government which is charged with the duty of providing for the lighting and electric service of municipal buildings in the City of New York, the Public Company supplying electric service in the region in which the building was located, and finally, a voluntary association interested in civic affairs.

These parties respectively are, The Borough of Manhattan, the Department of Water Supply, Gas and Electricity, the New York Edison Company, and the Bureau of Municipal Research.

While this test dealt with a technical subject, and was conducted with a considerable degree of scientific detail, it is nevertheless a matter in which the broad principles involved can be grasped by the non-technical reader if divested of unnecessary scientific terms and descriptions. The details with which it dealt, the main results attained, and the broad effects of the facts which were brought to light, are matters of interest to the intelligent citizen, and directly affect to a very substantial extent the operation of municipal affairs and the expenditures of taxpayers' moneys

The average property owner and taxpayer is usually unable to follow the operations of the municipal authorities in the disbursement of the funds which he and his fellow contributors provide, and is often unable to account for the inexplicable extent of expense involved in the conduct of municipal undertakings.

If light can be obtained upon one phase of this subject, such as the cost of operation of public buildings, the taxpayer may obtain therefrom some insight into the general methods which are being followed by those whom he has helped to elect to public office, and one of whose most important duties is the honest and economic administration of the funds provided by the taxpayer for the conduct of municipal affairs.

Owners of property also should take a particular interest in any matter which is concerned with the operation and expense of buildings, in which their pockets are directly interested.

The expense of maintaining city buildings is a subject that directly appeals to the general body of taxpayers, the majority of whom are owners of real estate. Under our form of city government they are the main contributors, through the value of their real estate, to the support of city institutions. In addition, the facts and figures brought out in a trial of such unusual extent, and dealing so comprehensively with certain interesting parts of the operation of buildings, affords information directly bearing upon similar operations under private ownership. In a word, therefore, the test of the plant in the Hall of Records from many points of view, possessed an extraordinary value, and warrants the attention of the public, especially of taxpayers, and of owners of real estate, so that the lessons which it affords may be directed to their own benefit and profit.

The mention previously made of the parties who united in the arrangements for conducting this trial, indicates the widespread interest in its results. Thus, the City officials recognized the great importance of ascertaining the most economical conditions under which municipal machinery could be expected to operate, and the best results obtainable therefrom, with a special view towards their guidance in the future establishment of plants of similar character, and the elimination of wastage and unnecessary expenses in the operation of the numerous installations of machinery already owned by the city.

The particular Department which is charged with the duty of providing electrical service for public buildings was peculiarly interested in ascertaining whether the policy of maintaining and operating such plants as that in this particular building, as well as those in many others, was productive of more economical results to the city than the alternative method of the purchase of steam for heating and of electricity for light and power from public sources, and also in determining the extent to which such services were economically utilized in city buildings; and what wastages and negligences in the use of these services had become established.

The Public Service Company was equally interested in the foregoing enquiries, on account of the bearing of the results upon the value to the city of their supply of energy, and they were also keenly alive to the importance of definite information upon the relative cost of their service in combination with that of other companies, such as the steam supply, and upon the opportunities afforded by some re-combination of apparatus or re-arrangement of methods which might be productive of equal or even superior results.

The Civic Association was interested in all the foregoing subjects from the point of view of the public interest in the disposition of its money in the form of capital investment in such machinery, and in the building space which it occupied, and was also concerned as to the current expenditures upon the services which it provided.

Outside of these parties, directly concerned in the conduct of the trial, were others who had a considerable indirect interest in the results. The great body of real-estate owners were doubtless those most directly interested in a pecuniary sense, and as has been previously said, they are those to whom the results ascertained present a very positive benefit.

The organizations interested in labor were greatly concerned in the subject, on account of the question of the labor involved in such municipal operations, and the rate of their remuneration. Engineers in general were interested in the problems of operation which the test was planned to dissect, and in the facts which were to be established over so unusually extended a period, enabling comparisons to be made of permanent technical value, and establishing standards from which the expense and value of operations of similar character might be measured.

The determination of the cost and the test of the operation of this plant is of permanent value in all these directions.

The trial extended over the entire year 1913, but the passage of time thereafter has not detracted from the practical value of the circumstances which were determined, which by ordinary processes of comparison can be utilized permanently as a basis for computation and information. Thus, the figures brought out in regard to the heating of these buildings during the particular year in which the test was made, are readily applicable to other seasons, and to other climatic conditions in different localities.

The facts in regard to the production and the use of the electricity made by this particular machinery can readily be applied to the operation of other machinery of a different character.

The analyses of the figures which were made after the trial by engineers representing different points of view of the elements under discussion, are of great value in bringing out the points upon which divergent opinions may be based, and in affording the means of analysis of their value and importance.

Altogether the results of this trial are of such general interest and importance that it would seem to be a very unwarrantable waste of valuable material if they should not be presented for general public reference in such a form as to be understandable and readily applicable to other circumstances, which so far has not been the case.

The author was one of those engineers charged

with the duty of conducting this trial, and was able to devote personal time and attention to observation of the results as they were progressively secured and developed. Being engaged in the designing and control of the operation of similar plants in many types of buildings in the same city in which the trial took place, the facts and details were subjected to analysis and comparison derived from experience with the same class of subjects, and it may be added that the points which have been determined have proved of extreme practical value in dealing with similar problems under other circumstances.

With these premises it may be hoped that the reader will appreciate the importance of a clear understanding of the facts which are now to be presented in as plain and untechnical a manner as such a subject can be set forth.

CHAPTER II.

THE BUILDING AND ITS NEIGHBORS.

IN the vicinity of the City Hall of New York City there are several municipal buildings which have been constructed from time to time, located in a very haphazard manner, partly on public park land and partly on sites recently purchased by the city. These buildings comprise the County Court, the City Court, the Hall of Records and the new Municipal Building, the last of which was, at the time of the test, in an incomplete condition, and was not included in the circumstances of the trial.

The other buildings had been connected by a system of underground piping for the conveyance of steam and water, and also with electric conductors both for light and for power purposes; all of these supplies being derived from the plant in the Hall of Records building.

The City Hall, which is the oldest building of the group, is the center of municipal government, being occupied by the Mayor and the President of the Board of Aldermen, and having two large halls used by the two representative bodies which divide between them the functions of government, the Board of Aldermen, and the Board of Estimate and Apportionment.

This little building, which is one of the architectural adornments of the city, has an area of 22,000 square feet, having a basement, two stories and an attic, the latter used by the Art Commission. The building was constructed between the years 1803 and 1812, at a total cost of one-half million of dollars or about 33 cents per cubic foot. It was originally heated by stoves, and has now been provided with a modern equipment of steam radiators. The gas chandeliers which long ago displaced the original oil lamps, are now serving a new lease of life as electric light fixtures.

The restoration of the interior of this building has only recently been completed, and was proceeding during a part of the period of the trial. This circumstance involved some irregularity in the service of heating and lighting, but the amount involved was not of an important extent.

A noticeable feature of the use of steam and electricity in this as in the other buildings is the wastefulness which prevails. Thus it was observed in the winter season prior to the trial, that the doors and windows of this building were left open in cold weather, wastefully condensing steam in the radiators, and it is, of course, a common experience to find electric lamps burning in broad daylight in this, as in other city buildings. The conduits providing these services connect the City Hall with the basement of the County Court Building, which stands immediately in the rear of the City Hall.

This is a rather widely known structure, which was erected under the administration of Mayor Tweed about 1872, and in the construction of which the improper diversion of public funds was exposed to a considerable extent. The total moneys expended in and about the construction of the building are alleged to have been in excess of ten millions of dollars, or at least $3.00 per cubic foot, a figure which, of course, is largely in excess of the real value of such a structure.

Compared with much more modern methods,

however, the extravagance does not seem to be so great, since wasteful and expensive methods still prevail, and extravagant plans and unnecessarily expensive materials are often found in municipal undertakings.

The County Court House has an area of about 37,000 square feet, and occupies a space of about 250 by 150 feet. It is lighted by electricity and is heated partly by steam radiators and partly by indirect heaters warming air which is blown into the court rooms. Until the year 1910, the heat was supplied by boilers located in the building, and a contract was then made with The New York Steam Company, which corporation extended one of its steam mains along Chambers Street and connected it to this building, supplying steam at about 100 pounds pressure per square inch. This steam was used to operate the pumping engines which supplied power to two hydraulic elevators, and the exhaust steam from these pumps was partly used in winter time to supply the radiators.

Prior to the trial of the Hall of Records plant this contract was abandoned and pipes were connected from the Hall of Records into the County Court House. At the same time the hydraulic elevators were changed to electrically-operated elevators, and the pumping engines were abandoned. These changes were all in the direction of economy if accompanied by a sufficient reduction in labor, but the methods of municipal operation are not contributory to economies of that character, and the men whose services were no longer required were in part retained in the building and in part transferred to other buildings, so that no actual saving resulted.

The use of this large building is quite irregular, as it is fully occupied only when the Supreme Courts are in session. On the ground floor there are offices which are in regular use daily throughout the year by the City Clerk and other departments, involving the maintenance of heat and light and elevator service at all times, although during the summer season other parts of the building are almost out of active service.

Close alongside of the County Court stands the City Court, sometimes described as the Municipal Court Building. This is an old-fashioned brick

The Group of City Buildings, City Hall Park.

structure of about 105 feet by 78 feet, which was erected in 1852, as a fire-house, and has been extensively altered and increased in height for use by the inferior or local trial courts.

It is provided with two hydraulic elevators, and is heated by steam radiators and lighted by electricity. For the purpose of operating the elevators hydraulic pressure pipes were laid underground from the Hall of Records Building.

The building is a great disfigurement of the appearance of the City Hall Park, and has long been destined to removal, which has only failed of accomplishment on account of the financial situation. Its operation by the services provided by the Hall of Records plant cannot therefore be considered a permanent condition.

The Hall of Records Building is by far the largest of the group of four structures described. It was built between the years 1901 and 1906 and is eight stories in height above the street, and has a basement below the street level, underneath which a sub-basement was provided for the purpose of housing the plant of machinery that was the subject of the trial.

The building occupies a site 194 feet by 151 feet, or an area of approximately 29,000 square feet. This was purchased by the city at a cost of $62 per square foot, and the building erected thereon, including the cost of the power plant, involved an expenditure of $5,979,343, or an approximate cost of $1.30 per cubic foot.

The building is ornate, its exterior overloaded with decorative features and statuary, and the whole structure disfigured by a disproportionate mansard roof, in which the two upper stories are inconveniently located. The interior is arranged around a central well or court, which is not of sufficient proportions to secure satisfactory interior lighting, and yet makes the means of access from one part of the building to the other extremely inconvenient. This condition is aggravated by the arrangement of all of the elevators on one side of the building.

The artificial light required in the interior of the building is excessive, and prior to the time of the trial the demand was increased by the nature of some of the fixtures in which the electric lamps were installed. These were extraordinarily elaborate and massive bronze chandeliers, the design of which appeared to have been dictated by an effort to detract as far as possible from the effect of light, and to cast as much shadow as possible therefrom.

The lighting of some of the offices, however, had been extensively modified and improved by temporary additions, but was utilized in the usual wasteful manner which has been described in connection with the same subject in the City Hall.

The building was, at the time of the trial, occupied by the Register of the County of New York, the Surrogates Courts and offices of the Commissioner of Records. On the ground floor the Tax Department of the city was at the time installed, and several floors in the upper part of the building were occupied by the Legal Department of the city. The two latter departments have since been removed to the Municipal Building, and other municipal organizations have been installed in their place.

The building at the time of the trial was already overcrowded, and the space available for the storage of the public documents for which the building was primarily intended, was found to be inadequate. Additional filing space was provided in the roof attic, and if it had not been for the presence of the machinery in the sub-basement, that large space would have been available for the same purpose.

The plant of machinery installed in this building was conceived in the same spirit of prodigality as the interior furnishings and exterior adornments of the structure. In so far as its proportions were concerned, it was approximately four times as large as was necessary, and its various appurtenances and the proportions of its piping were all upon the same scale of liberality. The general result of overproportions in apparatus of this character is naturally in the direction of inefficiency of the machinery which is too large for its duty, and is thus operated at only a small portion of its capacity, and in boilers, engines, piping and pumps, this always results in some loss of efficiency.

This overproportion of apparatus, however, is a not uncommon feature in municipal installations, which are usually large for the work that they may have to do. This effect is brought about by the natural conservatism of city employees and their desire to protect their own actions from failure by economy, and the further fact that they are not influenced by considerations of economy in the use of capital for such purposes.

The machinery is of excellent design and first-class material, and has been maintained in good operating condition. The circumstance of its over-proportion for the service of the particular building in which it was installed was of advantage in affording a capacity for providing the services required by the three exterior buildings, connected to the plant.

The result of this addition was an increase of the electrical service by fully 50 per cent., and an addition to the work of heating in winter of 175 per cent. Under these conditions the plant was operated at a desirably economical capacity, and still retained a sufficient excess to provide for emergencies in case of failure of any part of the apparatus.

The space which the machinery occupied was nearly the whole of the sub-basement. This space, about 14 feet in height, had been excavated and constructed at a large proportion of the expense of the erection of the building, and answers no other useful purpose than to house this machinery, except to afford a small space for the storage of disused interior furnishings.

The machinery space is very similar to other installations in the lower parts of modern buildings, and consists of a boiler room, other rooms devoted to the electrical generating engines and dynamos, to hydraulic elevator pumping machines, and other space is filled with small auxiliary appliances, chiefly in the form of pumps for sundry auxiliary services.

CHAPTER III.

THE MECHANICAL PLANT AND ITS DUTIES.

THE mechanical equipment in the Hall of Records building was planned for the purpose of providing that building with heating, lighting, elevator and water services, and also with a system of ventilation. The most essential element in such operations is that of raising steam, and this is effected in the Hall of Records, as in many other high-class buildings, by the use of high-pressure boilers.

At the time of the installation of these boilers there was no available public supply of steam, but at a later date the New York Steam Company's mains were extended to the immediate vicinity of the building, and thereafter the value of these boilers as an investment by the city for the purpose of generating steam, was limited by the price at which steam could be purchased from the mains in the street, which was at the time available at the rate of 42 cents per 1000 pounds weight of steam.

The capacity of the boilers which were installed was very considerably in excess of any probable requirements of this building alone, especially as such boilers may under proper conditions be expected to develop a larger capacity than that at which they are rated. It is not uncommon in boilers of this character to obtain as much as one-third more steam from a given capacity than the designed type, and this can be done also without loss of efficiency.

The plant consists of five separate boilers, three of which were of 206 horsepower, and two each of 238 horsepower, aggregating nearly 1100 horsepower. If a boiler plant had been installed for the sole purpose of heating these buildings, any two of these boilers would have afforded a very ample capacity. The additional boilers may be, therefore, regarded as having been installed for the expected additional demands to be created by operating electric generating engines and power pumps for the elevator service.

The boilers as originally used, were supplied

with coal of a large and expensive grade, which would burn freely with the draft afforded by the chimney. This chimney is constructed of steel plate and extends up through a well or shaft in the interior of the building to a point about 16 feet above the mansard roof.

Burning fuel of this grade under such conditions of draft, is, however, an expensive process, and like many other similar plants, a re-arrangement was made by the city engineers, by which a cheaper and smaller grade of coal could be employed. In order to burn such small coal, it is necessary to increase the draft, because the air must be drawn or pushed through a bed of coal very closely packed together. Therefore, a blower or fan was provided, operated by a steam-engine, which forced the air underneath the grates.

The use of this apparatus afforded the means of using coal of the grade known as No. 1 Buckwheat, which is the larger of the three smallest grades of coal, obtained by screening the waste or broken coal that used to be thrown away in the anthracite region, but is now utilized in increasing quantities and at gradually increasing prices.

During the term of the trial an effort was made to use a mixture of "soft" or bituminous coal with this anthracite, which can be economically done, provided that no objectionable smoke is generated thereby. It soon became apparent that the smoke was excessive, and public attention was drawn to it by the newspapers, so that a change was made and thereafter the trial was conducted with the use of anthracite coal, the average price of fuel during the entire year having been $2.90 per ton. The expense for fuel was $15,296.

It is scarcely necessary to say that since the year of the test progressive increases in the cost of production, and especially in the transportation of coal have largely enhanced its cost, so that at the present time the same grade of fuel is costing more than double the amount above mentioned. Such an increase would add over 50 per cent. to the cost of steam production, and would have increased by nearly 25 per cent. the total cost of operating the entire plant.

The supply of fuel to these boilers is provided in storage vaults located under the sidewalk, having a total capacity of 500 tons. This large capacity is a source of economy in the purchase of fuel, as the coal may be bought in large quantities at a lower price than it can be secured in smaller quantities, and the capacity of a coal barge is about 400 tons.

During the trial the fuel was drawn from these vaults through an iron door which allowed the coal to run into a wagon by which it was transported into the fire-room. On its way to the firing floor, the wagon passed over a weighing machine so that every wagon load of coal was weighed and the amount of fuel was thus accurately recorded. The arrangement is good, in that it relieved the firemen of much of the labor of lifting coal into the furnace doors.

The ashes are taken away from the boilers by being raked out onto the floor and shoveled into cans which were then wheeled to an hydraulic elevator situated under the sidewalk, by which the cans are raised to the street and are there lifted and dumped into wagons.

The boilers are supplied with water through pipes which are connected with pumps, known as feed-pumps, and on the way the water passed through a meter which accurately recorded upon a chart the rate at which the water passed through, at any given time, to the boilers. This water is drawn from several sources, the chief of which is, of course, the public mains. But this water, which is cold, is economically mixed with hot water derived from the condensed steam collected from different parts of the machinery and piping, and in particular during the winter time, with the water coming back from the heating radiators in the several buildings.

In the operation of a plant of this kind, however, there is never a complete return of all the water that has been made into steam, because some of the water becomes greasy and must, therefore, be discharged to the sewers, as the presence of oil in the boilers would injure their operation.

From the boilers the steam is directed into steam-pipes which in this plant extended around the machinery spaces in the form of a rectangular ring, the object of which was to provide for security in case of leakage or failure of one part of the piping, which while very desirable from the point of view of security, is not a source of economy. The size of the pipes was proportioned with the same liberality as the other parts of the apparatus. Such large and extended pipes containing steam act as radiators, and thus a considerable proportion of the steam traveling in the piping to the machines is condensed before it reaches the machines.

One of the most interesting facts ascertained by the trial was the extent of this loss by condensation, which for the first time afforded very definite information as to the amount condensed under different conditions of operation.

It was found that over the whole year of the trial more than 10 per cent. (10%) of all the steam which was made, was condensed in the piping before it reached the machinery. It would have required a boiler of 33 horsepower working at full capacity every hour of the year to produce this steam, and would have consumed 550 tons of coal. The condition is not unusual, and in fact is often exceeded in other plants. When the demand for steam is variable the condensation actually increases in the piping as the total steam used falls off, and in the Hall of Records there were times when the proportion rose to over 15 per cent.

It was greater in volume in July than in February. Comparing two such conditions in two weeks:

> February 24-March 2. Total steam made was 2,277,741 pounds and the steam condensed in the piping was 168,191 pounds.
>
> In the week, June 30 to July 6, the total steam was only 1,280,000 pounds, or about one-half, yet the condensation was 199,960 pounds.

The machinery to which the steam is supplied and which it is utilized to operate, consisted of three parts—the electric engines, the elevator engines, and certain auxiliary or minor pumps and engines.

The electric engines are installed in the center of the sub-basement space in a large and lofty engine-room, which is ventilated by fans blowing air in, and other fans drawing air out, in order to reduce the temperature in the room. The engines were four in number, each being provided with an electrical generator or dynamo, and any one or two could be independently operated, or all could be operated in unison.

The engines are of a high class of manufacture, two of them being of what is known as the Corliss type, of 240 horsepower each, and having generators with a capacity of 150 kilowatts. The smaller were respectively 160 horsepower and 100 kilowatts, and 80 horsepower, connected to a 50-kilowatt generator.

The electric energy generated in these machines is first led to a switchboard where any part of it may be diverted to one or other purpose. Part of the energy is used for electric lighting and part for the operation of power motors, of which there are no less than nineteen in the Hall of Records building alone. These motors operated ventilating fans and water pumps, and one of them was used to operate another small generator which produced electric energy at a somewhat higher pressure, and delivered the current into a storage battery which occupied a separate room. This storage battery is practically a reserve of electric power, which can be used at any time when the engines are not in operation, but, of course, only to a limited extent. The intention of its installation, which involved an expense of several thousand dollars, was to use this stored energy during the night time, to maintain lights in the corridors of the building when the engines should be closed down.

At the time of the trial, however, it had become the practice of these buildings to operate one of the engines all night, because under city methods of the employment of labor, it was necessary to keep an engineer on duty at night, whose services could as well be utilized in operating the engine. The battery, therefore, was practically out of service, and only sufficient use was made of it to keep it in operating condition.

The next large part of the machinery is that which is devoted to operating the hydraulic elevators. This consists of three large duplex steam pumps, arranged in a separate room and so proportioned that any two would suffice to maintain the hydraulic elevators in full operation. The water which is pumped by these engines, is reused, passing under pressure to the elevators and being discharged by them to a tank from whence the pumps again draw it. In order to maintain a regular pressure the water is pumped into large steel drums or cylinders, in which air is also pumped by two steam-driven air-compressors. Thus the air acts as an elastic cushion and prevents the shocks which would otherwise be felt in the elevators by the reversals of the pumps.

At the time these elevators and pumping machines were installed, considerable developments were taking place in the improvement of electrically-operated elevators, the use of which in this building would have rendered unnecessary the installation of these pumps, tanks and drums. Since the time of their installation successive improvements in the electric type of elevator have rendered it much superior to the hydraulic system, both in operating convenience and in economy of cost.

The effective period of life of such elevator plants has been shown by the course of events in other buildings to be about fifteen years, and not to exceed twenty years. The time is not far distant, therefore, when under natural conditions these elevators would be removed or remodeled into a system of electric operation, as has been done in many other buildings. Under such circumstances it would then follow that the operation of the elevators could be effected by using electric energy purchased from the public service, which of course could not be done so long as they are of the hydraulic form.

These remarks apply also to the elevators in the City Court Building, which together with the building in which they stand, are probably doomed ere long to disappear.

The elevators in the County Court Building had already been removed and replaced by electric machines, at the time of the trial, but these machines were necessarily operated by electricity purchased from the system of the New York Edison Company.

The next service provided by the machinery in the Hall of Records plant is that connected with water supply in several forms. Water for sanitary purposes and for the use of machinery is brought into the building from city mains from three points. Owing to the deficiency of pressure which has prevailed, especially in the lower part of the city, water from the mains will not rise in buildings much above the second story during the hours of the day when the greatest demand is made upon the city supply. This situation, of course, is now being modified and improved by the introduction of the Catskill water system, which reaches the city under a greater pressure.

During the trial, therefore, it was necessary to pump this water up to large tanks situated in the roof attic of the Hall of Records, involving the pumping of the water up to a pressure of ninety pounds per square inch. On its way into the building the city water passes through filters and is thence connected to the pumps. It is directed partly to the tanks in the roof and partly to other purposes.

The water supply in the other buildings, each of which is provided with its own tank, was connected by pipes to the tanks in the roof of the Hall of Records, so that all the water for all the buildings was pumped in the Hall of Records. Part of this water was heated for sanitary purposes, passing down through a pipe into a large heater provided with steam pipes inside, in which the water is brought to a high temperature and led by piping to the hand basins in the several buildings.

The circulation of this hot water through such long pipes was defective and it was, therefore, necessary to have a special pump operated by an electric motor, which forces the hot water around through the piping. Another portion of the water supply was used for drinking purposes, and in accordance with an old idea that prevailed at the time of its installation, it was considered desirable to cool this water in summer time, and in the Hall of Records a refrigerating apparatus was installed of the form known as the absorption type, which reduces the temperature of the water from the prevailing degree in the mains to approximately 45 degrees at the faucets in the Hall of Records building.

This apparatus was, of course, in use only a part of the year, and it involved the operation of an ammonia pump and a water circulating pump both of which were driven by steam.

Other large pumps were provided for fire purposes, which naturally are not in use excepting in such an emergency as a conflagration.

Another phase of the water service is that of discharging from the lowest part of the building the drainage of sanitary fixtures located below the level of the sewer. It will seem strange to readers who are not acquainted with conditions in New York City, to learn that such a necessity exists. The sewer system of the city, however, was laid out upon an old-fashioned idea that the depth of the cellars of buildings would not exceed a single basement and, consequently, the sewers are at a fairly uniform depth of only 13 feet below the level of the street. When, therefore, a sub-basement is constructed, as has very frequently been done, any water or drainage collecting in that lower level must be lifted by pumps to the sewer. Thus in the Hall of Records it was necessary to install two large sewage-ejectors, which, by means of compressed air produced by two steam-driven air-compressors, forced the drainage from the sanitary fixtures in the lower part of the building, up to the sewer. In addition, in case of the breakage of a water main, or the failure of steam piping, it is necessary to provide means to discharge a flood of water, and two large steam pumps were installed in the Hall of Records for this purpose, these being, of course, very rarely in operation.

Finally, the discharge of the waste waters of the machinery plant required the operation of steam pumps to lift this foul and greasy water, together with the water blown down from the boilers, up to the level of the sewer. It will be observed that these appliances were all involved in the construction of this sub-basement, the purpose of which was the providing of space for the operation of the machinery plant, consequently they become burdens upon the cost of operation of that plant, or from another point of view, are an expense attached to the value of the sub-basement space.

The same remarks apply to the ventilating apparatus rendered necessary by the operation of machinery, in order to keep the temperature down to a point at which workmen can be expected to conduct their labor. This naturally becomes a charge upon the operation of that ma-

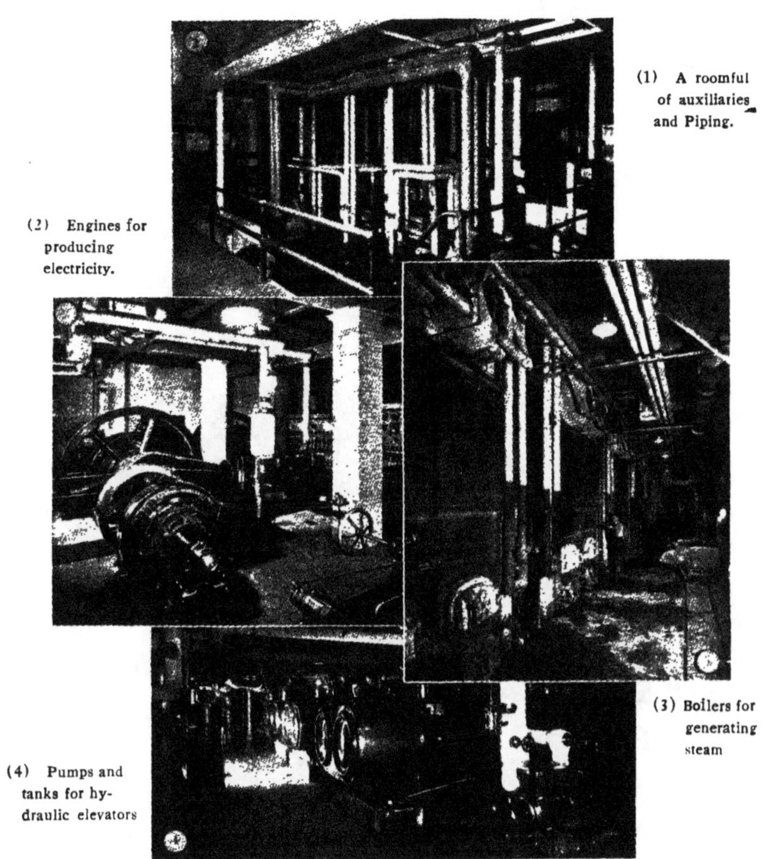

The Taxpayers' Investment in Machinery.

chinery, and in the case of the Hall of Records, the test determined the extent of this expense.

The apparatus further included the blowers or fans which afforded ventilation to the upper portion of the Hall of Records building, and this consisted of two chambers in the sub-basement in which these blowers were located, and large openings by which air could be drawn down into the building from the street level, warmed in winter time and discharged by the blowers up flues or stacks into the various offices. These blowers were rotated by electric motors receiving their energy from the electric generating engines. The arrangements in this regard were neither wholly effective, nor satisfactory in operation. The draft set up by the operation of these fans was objected to by the occupants of the offices. Moreover, the dust brought in with the air which was drawn from the sidewalk was excessive, and involved continual care in screening, which at the best, was not entirely effective. Thus, especially in the winter time, the fans were not utilized and air was allowed to find its way up in the shafts by natural draft.

This modification of the interior arrangements, of course, reduced the amount of work which had to be done by the electric plant, and the boilers.

The movement of the air thus described, was effected by keeping other fans in constant operation in the attic of the building, these fans having been installed for the purpose of drawing air out of the offices and discharging it to the atmosphere. The effect of stopping the blowers which were intended to force the same air into the offices was to produce a partial vacuum in the office-spaces, the tendency of which was to draw air in undue quantities through every crack and crevice in the windows and through the doorways of the offices.

This condition in the City of New York is productive of the introduction of dust in excessive quantities into a building, and in a structure devoted to the purposes of the Hall of Records the effect was much to the disadvantage of the occupants, and of the valuable documents which they have to preserve and handle. It also increased the work required to be done by the vacuum-cleaning apparatus. The difficulty is increased by the immense quantities of fine ash and cinder which are constantly emitted from the chimneys of power and heating plants in this city. The weight of such deposits varies from one to two per cent. of all the fuel burned, and is estimated to amount to about five hundred tons of ash per annum spread over each square mile of the city's area.

The final feature of the machinery plant consists of those small auxiliary apparatus which were connected with the minor functions of the steam machinery. These consisted of the feed-pumps to which reference has been made, which supply the water to the boilers, the drainage pumps which drew the water condensed in the heating radiators from the piping, and again discharged it to the boilers, and other pumps which received the condensed water out of the steam piping and discharged it to the boilers. All of these, it will be readily seen, were appliances the object and purpose of which it is to make practicable the operation of the steam boilers and the engines which they operate, and these are, therefore, as far as their cost is concerned, a burden upon the services provided by the plant.

The division or the proportion of the burden of expense of operating these appliances between the several services provided by the plant was one of the interesting studies developed by the trial. Some idea of the complexity of these minor appliances may be gained by a statement of their number. Of steam operated pumps, chiefly of the duplex pattern, there are twenty-one besides the elevator pressure pumps. There are seven air-compressors, six of which are driven by steam. Nineteen electrical motors of various sizes up to 40 horsepower operate fans and sundry other appliances. There are sixteen closed tanks and seven open tanks, besides feed-heaters, purifiers, separators and the accompanying maze of piping and valves.

CHAPTER IV.

THE TEST AND ITS METHODS.

THE task which was assumed by the Board of Engineers charged with the duty of carrying out the trial, consisted in establishing the entire cost of the installation of the machinery and the full expenses of its operation. The ascertainment of the money directly expended upon operation was a comparatively simple process, inasmuch as it involved only the maintenance of accounts by the Department, of its disbursements for labor, fuel, repairs and supplies. In connection with this matter, which involved the separation of accounting details on the books of the City Department, a committee of accountants was selected by the three parties in interest, and these gentlemen proceeded to establish a system of Cost Accounts, which were used in the Department during the year of the trial, in which were entered these separated expenditures.

The Board of Engineers also appointed a committee which went to work to ascertain from past records the amount of money that had been invested in the plant, and also to decide what was the rate of interest that the city was paying upon the money borrowed for the purpose of building the Hall of Records and its equipment. The Committee also decided the proper rate to set aside for the depreciation or annual reduction in value of the machinery as it became antiquated.

These facts and figures afforded the cost of the plant as a whole, and it then became necessary to divide the cost between the services which were provided by the machinery.

Inasmuch as the particular point at issue was the question as to whether the plant could produce steam cheaper than it could be bought, or electricity cheaper than it could be purchased, it was evident that the trial must involve complete measurements of the quantities of steam that were generated and the quantities of electricity that were produced, together with determinations of the uses to which these were put, so as to decide the expense of each of these services.

Moreover, a divergence of opinion had existed

as to the amount of steam which was really involved in the heating of such buildings as those included in the trial, and some engineers were in the habit of estimating upon quantities of steam required for the heating in the climate of New York, which other engineers maintained were largely in excess of those actually required.

The advocates of the operation of these small power plants in municipal and private buildings have maintained that the amount of steam required in heating buildings could usually be provided by the waste steam passing away from the engines after it had been used to generate electric energy. Other observers pointed out that this would be all very well if the generating of electric energy took place exactly at the same time and exactly in the same quantity as was required to provide the amount of steam necessary under all the variations of our extremely erratic climate in New York, a condition entirely impracticable of attainment.

The advocates of electrical elevator operation believed that the use of elevators operated by electric energy was more economical than that of the hydraulic apparatus, and this involved the decision of the number of trips made by the elevators and the amount of steam and other expenses connected with their operation.

No one, prior to this trial, had ever thought it worth while to consider what it cost to pump city water for use in the sanitary fixtures, and comparatively few people had any idea of the expense involved in cooling water for drinking purposes. All of these features and a number of other important points involved a close and detailed measurement and record of the several operations conducted in the plant from hour to hour, requiring provisions for a large number of measuring appliances, some of which were quite complex and expensive, and others of simple and usual character, together with arrangements for a corps of observers, who should maintain strict watch over all the details of the operation, without the least cessation of their attention.

First, a study was required of all the piping and valves controlling the flow of steam and water, and the switches controlling electric energy, and these were made the subject of plans by which the movement of the steam and water could be traced in advance. At the necessary points the pipes were then cut open and meters placed in them, so as to record the flow of the fluid, and electrical meters were placed on every line of conductors from each generator, and the storage battery, and upon each line, extending to the purpose of power or lighting. Most of these appliances were self-recording devices which were kept under lock and key, in charge of the particular observer on duty.

The observations further extended to the weighing of all the fuel and all the ashes in and out of the boiler-room, also the recording of the variations in pressure of the steam, and the observation of exterior temperatures, of the barometer and wind pressure, all of which have some effect upon chimney draft and upon the heating of the buildings.

The extent of this work may be judged from the fact that upwards of twenty meters were required to record only the movement of water, and nine meters were installed upon steam supply lines, with fourteen special electrical meters, besides those in regular use on the switchboard. The records from all these meters were removed daily and filed in an office provided for the purpose by the Department of Public Works, in the care of an official.

Visual observations and readings of thermometers were made every two hours, day and night, and these were also filed daily.

The work of observation required the attendance of two observers on duty at all times, and a corps of assistants was constantly employed in checking, comparing and recording, and computing the results of the observations. A weekly summary of the daily observation was prepared and submitted to the Board of Engineers, who thus were made acquainted with the progress of the observations and the progressive effects which they disclosed.

To conduct the trial and direct the preparation of the figures and results, the Board selected Dr. Herman Diederichs, of Cornell University, under whose charge the observers at the plant and the computing and recording staff at Cornell University were engaged.

One of the most anxious and difficult tasks entrusted to the test engineers was the maintenance of the accuracy of the devices used in measurements. This work involved constant observation, in order to discover any tendency towards error on the part of the different devices, and required the testing of such devices from time to time, to ascertain their accuracy. In the case of certain meters of similar size on similar apparatus, where divergencies appeared that were not readily explainable, the devices were removed and placed in each other's positions, and then again replaced. Others were taken to the laboratory and tested, all of which involved considerable labor and care, but rendered the general results of satisfactorily dependable character.

With all preparations made, the observations and records commenced at midnight on December 15, 1912, and for the period of an entire year thereafter, the record of two-hourly observations, supplemented by the automatic records of the various meters, decided the following details of the operation of this plant:

The weight of the coal and of the ashes.

The character and heating value of the coal.

The moisture in the coal.

The measurements of the water used by the boilers and its temperature.

The volume of the steam discharged from the boilers.

Its pressure and the amount of moisture it contained.

The condensed water drawn from the steam-pipes.

The distribution of the steam to the various apparatus and the proportional amount of the steam used by each of the total. (This item included the amount of steam taken direct for the purpose of heating buildings in the winter season, which was a separate measurement from the amount used for the heating purposes, drawn from the waste or exhausted steam of the engines and pumps.)

The steam used by the engines for generating electricity.

The steam used by the pumps for operating the elevators.

The steam used for the smaller services such as pumping sewage, heating hot-water, operating the refrigerating plant, and pumping house-water.

The steam used by the feed-pumps, drip-pumps, return-pumps, and all auxiliaries used in and about the steam power plant.

In addition to the foregoing, observations were taken of the travel of the elevator-cars in all the buildings, the amount of water used in the buildings for sanitary purposes, the amount of iced-water cooled and used, and many other details, such as tests of the gases discharged from the boiler furnaces, observations of the smoke emitted from the chimney, determination of the amount of electricity required to operate fans to ventilate the boiler and engine rooms, and the steam involved in operating apparatus necessary to circulate the heating-steam in the radiators of the buildings, with other records such as the wastages of steam drawn from the engines and pumps and discharged to the sewer.

In connection with all this work the division of the cost of labor and the materials required by different parts of the apparatus was of great importance, and daily reports were made upon these items.

The most important part of the work was the determination of the quantity of steam used by each apparatus, and the proportion of exhausted steam returned to heating purposes. These facts were ascertained and presented in the following:

SUMMARY OF RESULTS OF THE TEST OBSERVATIONS FOR 52 WEEKS, BY THE TEST ENGINEER

Live Steam Distribution

	Lbs.	Per Cent
Total steam generated	81,656,863	
Absorbed by condensation losses in piping, etc. (clean drips)	8,736,125	
Proportion of total steam generated		10.7
Absorbed by boiler auxiliaries (feed pumps, etc.)	8,416,435	
Proportion of total steam generated		10.32
Absorbed by exhaust heating appliances.	3,822,004	
Proportion of total steam generated		4.68
Absorbed by electric power-engines...	36,275,352	
Proportion of total steam generated		44.48
Absorbed by elevator-pumps	12,509,868	
Proportion of total steam generated		15.33
Absorbed by minor services (sewage pumping, house pumping and refrigeration)	8,694,615	
Proportion of total steam generated		10.64
Absorbed by live steam to heating services	3,202,897	
Proportion of total steam generated		3.92
Diverted in auxiliaries and drips.....	19,205,044	
Proportion of total steam usefully utilized		76.40
Total steam used in house heating and hot-water heating	31,396,978	
Proportion of total steam generated		38.40
Exhausted steam used in above.......	27,609,926	
Proportion of total heating......		87.95
Proportion of total steam generated		33.8

Live steam used in heating and hot water	3,537,038	
Proportion of total steam generated		4.34
Proportion of total heating steam		11.28
Steam delivered to electric engines....	36,275,352	
Proportion of total steam generated		44.48
Contribution of engines to heating	11,250,956	
Proportion of total steam delivered to engines		35.87
Total steam absorbed by exhaust-heating, vacuum and drip appliances...	3,822,004	
Proportion of total steam generated		4.68
Proportion of exhaust steam utilized		14.06
Exhaust Steam Distribution		
Total exhaust available for heating work (house heating, hot-water heating and refrigeration plant)	52,587,835	
Exhaust steam utilized in heating work	28,484,000	
Proportion of total exhaust steam available		54.1
Proportion of total steam generated		34.7
Exhaust steam discharged to atmosphere	23,852,128	
Proportion of total exhaust steam available		45.4
Proportion of total steam generated		29.2

The operation of the machinery was maintained by the regular staff of fourteen men, under the direction of the chief engineer of the building. The duties of these men and the details of the maintenance and supply of the plant were superintended by a mechanical engineer detailed for the duty by the Department of Public Works.

The general effect of the trial was to provide the means of determining the working conditions of the various services over the period of a year, with particular information as to its operation at any time during that period, thus affording a complete picture of the variations and conditions at different seasons and under climatic changes, and with the variations which are due to the working and non-working hours of the building.

After the completion of the test on December 15, 1913, a period of about three months was occupied in the re-checking and tabulating of the whole of the facts and figures for the presentation of the Test Engineer's report, which was extremely comprehensive. Thereafter, considerable time was occupied by the members of the Board of Engineers in an individual study of the results, and the preparation of their opinions thereon.

Some further delay resulted on account of a divergence of opinion as to the bearing of the conclusions of the trial upon the question of alternative operation of these buildings for another year by means of purchased steam and electricity, as had been agreed upon. This and other delays resulted in retarding the presentation of the facts until 1916, when a brief statement of the main financial facts was published by the Author. This was followed in the year 1917, by a very voluminous compilation of all the material connected with the inception of the trial, its conduct, the reported details, and a large amount of correspondence and briefs, which was published by the Bureau of Municipal Research, and failed to present the facts and figures in summarized form.

It is in the hope that a review of the essential features of the trial, and of its unquestionable results, may be found more readily accessible and comprehensible, that this story of the test has been prepared.

CHAPTER V.

THE FULL COST OF THE PLANT.

THE widest feature of the subject dealt with by the trial at the Hall of Records, is the question of what desirability or advantage results from an investment in such plants of machinery, and the bearing which the result would have upon further investments of a similar character.

The taxpayer would naturally be less interested in the question of the continued operation of a plant already installed than he would be in the question of the advisability of having expended his funds in its establishment. The results provided by the plant must be weighed against the capital expense which was involved, and if those results could be obtained with equal or less cost by some other means which do not involve the use of the capital provided by the community, then there is no good reason why the expenditure should have been made, or why a policy of similar expenditures should be continued.

The expense involved in purchasing and installing machinery under municipal conditions is always greater than that of similar operations under private control, and the Hall of Records was no exception to this condition. The money which has thus to be expended upon machinery by the municipality, is not available, as in the case of private enterprises, from the usual source of free capital, but it must be borrowed upon the credit of the community. That credit is of course, limited by the law, and in the State of New York it is restricted to one-tenth of the current assessed value of all property within the municipal limits.

Another effective limitation is that of the actual credit of the municipality, which is reflected in the rate of interest, demanded by financial interests and by investors, upon loans made to the city. Thus the larger the indebtedness of the city grows the higher the rate of interest required, and if the money thus expended is fruitlessly invested, the result is that money required for necessary purposes, or for more productive purposes, is not only more difficult to obtain, but is more costly when secured.

In addition, all investments in machinery are short-lived. Complex moving machinery in isolated plants, such as that in the Hall of Records, does not have a useful existence of more than at most, twenty-five years. But in order to secure money at reasonable rates it was found necessary at the time of the installation of this plant to borrow the funds for the purpose of the construction of the building and its equipment by the issuance of bonds having a life of forty years.

The issuance of long-term bonds of this character bearing interest at the rate of 3½ per cent for forty years, involves the eventual payment for each one hundred dollars, of $140 in interest and $47 in sinking fund.

In expending money thus borrowed upon machinery it will be evident that at the time the machinery has become worn out or has to be replaced, the indebtedness will not have been paid, so that the interest and sinking fund charges will continue for another period of years, and meantime, a second issue of bonds must be made to replace the superannuated machinery.

The borrowing of money for such purposes by a municipality, always involves the return of the capital at the end of the stipulated term of the bond. The only way in which this can be accomplished is to set aside each year, out of the taxes paid by the taxpayers, a sum which invested at compound interest will, at the end of the term of the bond, repay the capital to the lenders. Thus in the end, the taxpayers pay every dollar of the expense involved, and in the case of machinery, they are for a great part of the term of the bond, paying for something that has passed out of existence.

The process is entirely different from private investment in the stock of commercial corporations, which is a continuing investment, not involving repayment at any particular date.

In the case of the Hall of Records the large sum of $5,979,000 was borrowed and expended upon the construction of the building, by the sale of corporate stock, the major portion of which matures in the year 1940. In addition to this,

$1,610,000 was borrowed to pay for the land occupied by the building, making the total expenditure nearly seven and a half millions of dollars, repayable in about thirty-five years.

Of this borrowed money, about $118,532 was expended on the steam and electric power plant, at a rate of interest averaging 3.48 per cent. The annual interest on the plant thus amounted to $4,125, to which the annual amortization charges or contribution from the taxes to repay the capital, added approximately $835 a year, making a total of $4,960.

A large sum of money was expended in providing the space in the sub-basement for housing the plant of machinery.

This space having a certain practical usefulness in a public building of this character, may be regarded as a proper expenditure if used for its particular purposes, but it must be evident that if it be occupied by machinery, then its use for practical purposes such as the storage of documents or goods, is rendered impracticable. As the city is compelled to pay rental for space in a number of buildings for just such purposes, and

The County Court House—Tweed's Monument.

loss which the city incurs by taking away private property which is paying taxes, and diverting it to public purposes which do not pay taxes. In the case of this building it was ascertained that the land and its improvements brought in to the city an annual income of $70,830, which of course ceased when the land was taken over by the city.

Another element which enters into the cost of the operation of all such municipal buildings and their equipment, is that of supervision, which is as necessary under municipal conditions as under private control, and is always relatively an expensive item. The cost of this duty in the case of city buildings is not difficult to ascertain. They are all placed under the direction and charge of a special bureau, known as the Bureau of Public Buildings, which is a part of the organization of the Department of Public Works.

The accounts for these Departments are all available, and the expenditures which are made therein upon salaries, office expenses and contingencies, are distributable over all the operations conducted by the Department.

It was found that these expenditures represented an addition of 6.38 per cent. to the cost of the upkeep or maintenance of all the public buildings and their equipment, and therefore, that this percentage should be added to the expense of the operation of the Hall of Records building and its machinery.

For the year of the trial it was found the total operating expense of the entire building, including the machinery, amounted to $60,572, and to this the expense of supervision added $3,852, of which six-tenths was required by the supervision of the machinery.

We now reach the last of the large items of standing cost in the operation of this public building, which is the somewhat difficult element known as depreciation. This is a financial provision which is often misunderstood, because the term is frequently misapplied to the physical failure of a building or of machinery, which is properly to be described as deterioration and obsolescence. This latter term describes the change which takes place in the commercial value of a building or a machine, due to changes in the use of either, or to the development of other methods of accomplishing the purposes for which the building or the machinery was erected. The provision for the result of these processes is what is known as depreciation.

As regards machinery, the circumstances which may have to be provided for are less readily foreseen and calculable than they may be in the case of a building. Thus a building may be useful for fifty to one hundred years, with reasonably extensive changes and alterations, although no one could say that at the end of a century it would still be as commercially valuable as it is at the present time. In the case of a municipal building, for instance, radical changes may take place in the convenient location of city government, or even in the character and the methods of that government, which cannot be foreseen.

In so far as its mechanical equipment is concerned, however, past experience indicates that within a period of less than a quarter of a century all machinery becomes antiquated and worn-out in vital parts, and should be removed and superseded by other and more modern apparatus. Thus the provision for the loss of this machinery, as well as some eventual decline in the usefulness of the building, should properly be made in any thorough study of the financial expenses involved in their construction and maintenance. This is the process known as depreciation, which is a financial system of providing in advance for loss of the value upon which the capital has been expended.

The subject is further misunderstood by those who represent that inasmuch as the original capital invested by a municipality in the form of borrowed money upon the building or plant, is to be repaid out of the taxes at the end of the term of the bonds, that the property then belongs entirely to the taxpayer, and has been paid for in full out of his pocket, and, therefore, no necessity exists for the establishment of a depreciation fund, which at that time or any other time shall be accumulated so as to replace the machinery or building, which has then been paid for.

This view is mistaken, because just as in the case of a privately-owned property provision for a depreciation fund is entirely necessary, the only difference between the two financial processes is, that the commercial or private investor has paid out the money for his building or his plant in cash at the time of the construction, while the taxpayer has borrowed the money and repaid it in annual contributions. At the end of a certain term the building and the machinery become useless and the private investor and the taxpayer finds that he has spent his money for something which represents now a decreased value. Where will the money come from that shall replace the building or the machinery?

Unless a fund has been accumulated out of the earnings of the building or the plant, or out of suitable contributions by the owner for the purpose, then another new borrowing has to be made to replace the building or its plant.

Thus the proper and careful consideration of the circumstances of the plant in this building as in all others, include a charge for depreciation.

There was no division of opinion between the engineers engaged in the test, upon this general subject. The proper rate to be applied was studied very carefully, and varying rates were applied to the machinery, having regard to the probable useful life of the various parts of the plant. In this way they arrived at rates which total $7,407, as representing the depreciation of the steam generating plant and machinery.

The foregoing considerations thus afforded the means of summarizing the total yearly cost of operating the Hall of Records building, which aggregated $521,000.

The ascertainment of these several features of expense now rendered it possible to proceed to the consideration of the total cost involved in the operation of the machinery as separated from the general operation of the building. The determination of the exact amounts expended on the plant during the year of the trial, was a sum of $39,311. This represented what is known as the operating costs of the machinery plant, and covered only the payments made for fuel, supplies, repairs, and labor in the engine and fire rooms.

There were a number of items to be added to this operating expense, in order to afford a complete statement of the total expenditures, in which the city was involved in installing and operating this plant.

Thus, the annual payments of interest were	$4,960
The depreciation upon the plant, added	7,407
The expense of the floor space was	7,828
The supervising expense was	2,520
And the operating disbursements, with some small items of unapportioned supplies	39,514
Making a total of	$62,229

This was the full expenditure of the taxpayers upon the operation of this plant for the year 1913. For this sum of money the results secured are the following services and their cost:

1. The heating during the winter season of the four buildings, including the heating of warm water supplied for sanitary purposes	$16,736
2. The generation of electric energy for lighting the buildings as follows:	
The Hall of Records	12,480
The County Court House	5,302
The City Hall	2,955
The City Court Building	1,375
The operation of the ventilating fans in the Hall of Records and sundry small power purposes	1,270
The operation of the storage battery....	1,004
The lighting of the machinery spaces and boiler room, and the operation of fans for ventilating those spaces	5,008
3. The operation of the elevators in the Hall of Records and the City Court Building.......	10,238
4. The pumping of a supply of city water to the several buildings for sanitary purposes	1,980

5. The cooling of drinking water for use in the Hall of Records 2,799
6. The pumping to the sewer of the drainage from sanitary fixtures in the sub-basement of Hall of Records 1,214

The costs thus ascertained represented the value of the net services secured by the operation of the plant. Those items which were merely contributory to the operation of the plant itself, are not useful products of the machinery. All that was obtained for the total sum expended was, in brief, the heating, the lighting of the several buildings, the partial ventilation of one building, the elevator service of two buildings, the provision of hot and cold water service for all buildings, and the cooling of the drinking water in one building.

In considering the division of expenses among these services, it will be evident that unless a fair distribution be arrived at, the cost of one may be unduly enhanced, while the cost of others is reduced.

It would be unproductive of a common-sense result to assume, for instance, that the electricity produced was without any cost, on account of part of its steam having been used in the heating

The costs of the services divided between the several buildings were as follows:

Building	Heating and Warm Water	Electric Lighting	Electric Power and Storage Battery	Lighting and Ventilating Machinery	Elevator Operation	Water Supply	Drinking Water	Pumping Sewage	Totals
Hall of Records	$10,042.10	$12,430.00	$2,200.00	$2,275.00	$8,190.66	$900.15	$2,798.93	$1,214.37	$40,051.21
County Court	3,816.40	5,302.00	......	1,550.00	......	613.14	......	......	11,281.54
City Hall	1,874.18	2,955.00	......	765.00	......	303.50	......	......	5,897.68
Municipal Court	1,004.14	1,373.43	......	410.00	2,047.66	163.54	......	......	4,998.77
	$16,736.82	$22,060.43	$2,200.00	$5,000.00	$10,238.32	$1,980.33	$2,798.93	$1,214.37	$62,229.20

of the building. The result of such an assumption would be to throw the burden of cost upon the process of heating to such an extent as to make it more expensive than it would have been if conducted by steam purchased from the Steam Company. Inversely, if it were assumed that the heating of the building cost nothing on account of its having been accomplished in great part by exhausted steam, then the effect would be to cast the burden of the greater part of the expenditures upon the production of electricity, or other services.

The test, however, established the relative quantities which were used for each service, of the fuel, the supplies, the repairs and the fire-room labor, so that the distribution of these expenses is readily made.

The test decided that the steam was used in the following proportions:

	Per Cent
Heating	40.57
Electricity	39.82
Elevators	12.05
Water Pumping	8.8
Cooling Water	2.85
Sewage Pumping	0.91

This disposed of the expenses of the production of steam, and left only the engine-room labor, supplies and repairs. In great part these items were distributed by the daily labor time sheets, and with small exceptions the department charged the cost of supplies and repairs to each service.

Thus, one-half of these items were decided in the following proportions, and the remainder, which was chiefly composed of the salary of the chief engineer and his three assistants, could be fairly apportioned in the same relation.

	Per Cent
Heating	30
Electricity	43.6
Elevators	16.3
Water Pumping	3.6
Cooling Water	4.4
Sewage Pumping	2.1

The expense of space, and the interest and depreciation charges could be readily dealt with by separating those relating to the steam plant from those of the machinery.

These fixed expenses connected with the steam plant are divided in proportion to the use of steam by each service.

Those relating to the machinery are directly proportioned to their size or the space occupied, their original cost, and their probable term of useful existence.

The total cost demonstrates that the production of steam and the generation of electricity and the other operations were conducted at an expense considerably in excess of what might be accomplished by other methods, combined with the purchase of electricity, even with the maintenance of the present fixed charges.

In point of fact, by the city having gone into the expense of installing this plant for these purposes, the operation was conducted at a loss, amounting during the year of trial to fully $30,000.

The result of the investigation of the expenses, whatever be the division of the cost to one or other of the services it supplied, could lead to no other general conclusion than that the investment of the city in this plant was a financial mistake, and that the policy of the establishment of such plants in municipal buildings is not justifiable from any financial standpoint.

Distribution of the total expense of installing and operating the power plant in the Hall of Records. Year of 1913.

Services provided by the plant		House heat and hot water	Electric light and power	Elevator operation	Water pumping	Cooling the drinking water	Pumping out sewage	Totals	Summary
Per cent of steam used		40.57	39.82	12.05	3.80	2.85	0.91	100.	
Cost of the production of steam	Coal	$6,203.59	$6,091.00	$1,822.00	$581.21	$459.58	$139.20	$15,296.58	Expense of production of steam $28,022.98 Per 1,000 lbs. produced, 34.4 cents
	Ash removal	495.00	485.80	147.00	46.36	34.72	11.10	1,219.98	
	Gas	102.25	100.30	30.36	9.68	7.42	2.29	252.30	
	Boiler repairs and supplies	378.10	370.80	111.60	35.30	27.99	8.46	932.25	
	Fire room labor	3,505.65	3,438.00	1,041.10	328.30	249.60	78.60	8,641.25	
	Production cost	10,684.59	10,485.90	3,152.06	1,000.85	779.31	239.65	26,342.36	
	Supervision	681.00	669.00	201.00	60.32	54.00	15.30	1,680.62	
Cost of operating the machines	Labor as apportioned	415.60	2,309.75	1,040.02	147.67	193.08	109.15	4,215.28	Expense of operating machinery, $14,011.22 Adds per 1,000 lbs. of steam, 17.3 cents
	Remainder on same basis	639.00	3,557.52	1,600.50	226.10	298.00	159.54	6,480.66	
	Supplies and repairs as apportioned	395.62	786.56	606.14	50.94	417.76	31.78	2,288.80	
	Remainder on same basis	31.60	63.20	49.10	4.15	33.48	4.95	186.48	
	Operation cost of machines	1,481.83	6,717.03	3,295.76	428.86	942.32	305.42	13,171.22	
	Supervision	94.40	427.50	210.50	27.30	61.30	19.00	840.00	
Total operating expenses		12,941.82	18,299.43	6,859.32	1,517.33	1,836.93	579.37	42,034.20	
Fixed or Overhead Expenses	Cost of space, boiler plant	1,495.00	1,466.00	422.00	140.00	104.00	57.00	3,684.00	Fixed Charges on boiler plant, $9,360.00, on machines, $10,835.00. Adds 50% to operating expense
	Cost of space, machines	....	2,636.00	1,196.00	....	156.00	156.00	4,144.00	
	Interest on boiler plant	962.00	943.00	285.00	90.00	68.00	27.00	2,375.00	
	Interest on machines	....	1,773.00	420.00	42.00	210.00	140.00	2,585.00	
	Depreciation on boiler plant	1,338.00	1,313.00	396.00	125.00	94.00	35.00	3,301.00	
	Depreciation on machines	....	2,830.00	660.00	66.00	330.00	220.00	4,106.00	
Total fixed charges		3,795.00	10,961.00	3,379.00	463.00	962.00	635.00	20,195.00	
Total annual expense		16,736.82	29,260.43	10,238.32	1,980.33	2,798.93	1,214.37	62,229.20	
		Per 1,000 lbs.	Per k.w. hr.	Per car mile	Per water h.p. hr.	Per 100 lbs. ice	Per 1,000 cu. ft.		
Unit Costs and Value	Quantity produced	50.4 cents	4.25 cents	62 cents	10.4 cents	33 cents	$15.60		Saving by purchased electricity and steam
	Quantity required to be bought	62 cents	5 cents	......	......	......	......	$47,406.20	$14,823.20
	Price of public service	40.4 cents	1.92 cents	18 cents	7 cents	17 cents	$9.40		

CHAPTER VI.

THE COST OF PERSONAL SERVICE.

INASMUCH as the city had already made the investment involved in the installation of this plant, and the fixed expenses in connection with the capital could not be brought to an end even if the plant should be put out of service or removed, the next question of interest is what economy might be found in the partial or complete substitution for the operation of the machinery of purchased supplies of steam or electric energy. Naturally, in connection with such a consideration, the largest element which would affect the results would be that of labor. This would be true to a great extent in any commercial plant, but it is particularly prominent in the case of municipal operations.

The cost of labor in this, as in other municipal plants of machinery, is relatively high. It was by far the largest single item of expense, being fifty per cent. of all the operating disbursements. In well managed private plants of a similar character the expense of the labor employed about the plant has been generally found to be about equal to the cost of fuel, prior to advances in price more recent than this test.

In some plants under private management of the size of that in the Hall of Records, labor would in the past have cost considerably less than fuel, but under municipal management it is always the case that the cost of labor has exceeded that of fuel, sometimes very largely.

Upon the prices of coal which prevailed until quite recently, the labor in city buildings will be found frequently to cost twice as much as the fuel. In point of fact, in the Hall of Records, prior to the investigation which resulted in the test the labor cost $28,000 a year, as against the fuel cost of nearly $14,000.

During the period of the trial, when every effort was made to economize, both in labor and in fuel, the former was reduced to $19,100 as against a fuel expense of $15,296.

The comparison may be more clearly understood by relating the cost of the labor employed in the fire-room in handling the fuel, to the weight of the fuel. In this building the cost of fire-room labor was $1.67 per ton of fuel handled. The average cost under private operations in well managed plants, varies from 55 cents to 85 cents, upon the same character of fuel and in the same locality as the Hall of Records.

It will, of course, be observed that the rate of pay of firemen under municipal operations, is much higher than that paid in private service. In the latter the average wages paid to firemen per diem vary from $2.15 to $2.50 for a day of ten or twelve hours, while in municipal service the firemen receive $3 a day for eight hours' labor. But even so, the amount of work accomplished by the firemen is much less than would be expected in the case of private operations.

During the conditions prevailing prior to the trial, it was observed that the firemen in this plant handled an average of only two tons of coal per day per man. As an illustration of the amount that may be handled by a less amount of labor under private management, one building may be mentioned in which the average amount handled per diem per man is three tons, and during short periods of very cold weather, it has not been uncommon to observe as much as a ton and even a ton and a half an hour, being handled by a single fireman, aided by a coal-passer, or helper.

The relation which the labor bears to the amount of fuel handled, of course, varies with the seasonal demands for the production of steam, and thus in summer time when the amount of fuel consumed is very greatly reduced, the same force of labor is, under municipal conditions, usually maintained, and the relative cost in the Hall of Records rose to no less than $2.40 per ton handled, nearly doubling the cost of the fuel.

A similar condition of expensive proportion of labor during the summer season is to be found in nearly all private power plants operating under conditions such as the service of office, business and mercantile buildings. The demand for electric light and for steam for heating purposes, of course, is greatly reduced, and yet the plant must be kept in operation, and the services of valuable men may have to be maintained, in order to insure their employment in the winter season. It follows, therefore, that labor is often most uneconomically employed under such circumstances during one-half of the year.

In comparing the cost of steam production by private installations with the cost of the supply of the New York Steam Company, it is, therefore, necessary to look into the conditions in the fire-room during the entire year, and in day and night operations.

The circumstances are less advantageous to the operation of a private plant under municipal management, on account of the legal restrictions upon the hours of labor. It was necessary in the Hall of Records, as in other municipal buildings, not only to employ three sets of firemen to complete the twenty-four hour service, but to maintain a licensed fireman on duty at night in the plant, whether there were any boilers in actual service or not. It thus came about in the Hall of Records, that during the summer nights, when the boilers were standing with banked fires, and rendering no service whatever, a gang of licensed firemen were maintained on duty. The same conditions apply to the labor employed on and about the machinery. These men are paid higher wages than are obtainable in private service, the assistant engineers receiving four dollars and fifty cents a day, and even oilers, an unskilled form of labor, receiving three dollars per diem.

All of these men work only for eight hours, and in their case also a licensed engineer is required to be on duty in any engine-room where high-pressure machinery is installed, whether the machinery is in operation or not.

During the summer season, with the employment of the storage battery in this plant, it was not necessary to keep any of the machinery in operation at night, yet a full staff, composing an engine and fire-room watch, was on duty every night.

It may well be imagined that organizations interested in labor conditions for the benefit of the skilled workmen, are actively interested in the continuance of the operation of such plants, as affording the means of employment to a number of their members. The effect of any economical changes are keenlv scanned by those thus interested, with a view to opposing any reduction of the labor which may be employed.

The staff had been reduced to fifteen men, and under ordinary circumstances would probably have included a greater number. Some complaint was made during the trial by the representatives of labor, that the number had been too much reduced in the effort to secure economy during the period of the trial.

Advocates of the maintenance of such plants, both in municipal and private operation, usually hinge their arguments for the continuance of the use of such machinery upon the necessity for the continued employment of labor, even if the operation of some part of the apparatus be dispensed with, by the purchase of steam or electricity, or refrigeration, or water, from other sources. Thus in the case of the Hall of Records the question as to the amount of labor that would necessarily have to be retained in employment upon the disuse of the electrical generating engines, or of the steam-generating boilers, became the subject of considerable discussion and of some divergence of views between those regard-

The City Hall of the City of New York.

ing the matter from a practical or commercial standpoint, and those viewing it from a departmental point, or from their interest in the operation of isolated plants.

The item of labor in the department was, therefore, that upon which the main contention centered.

Only part of this labor was charged daily in the Department accounts to the separate services, leaving about one-third undistributed.

The effect on unit prices, of leaving this undistributed expense out of consideration, would be as follows:

Service		Per Cent of Total Cost	Unit Cost Affected
Heating .	$639	4	2 cents per 1000 lbs.
Electricity	3557	12	5/10 of a cent per k.w.hr.
Elevators .	1600	15	9 cents per car mile
Water . . .	226	11	6 cents per 1000 cu. ft.
Cooling . .	298	10.7	3½ cents per 100 lbs. ice
Sewage . .	159	13	$2.00 per 1000 cu. ft.

And it still leaves the operating cost only of electricity, exclusive of any fixed charges and space, at 2 cents per k.w. hr.

The minimum reduction which was proposed by those whose view it was that the operation of the plant should be maintained, provided that upon the discontinuance of the electrical generating apparatus, the services of two oilers could be dispensed with, and the fire-room staff reduced to three men for the two hundred and fifteen days of the summer season. This made a total reduction in expenses of $4,125, out of $19,102, or about 20 per cent.

A closer investigation of the actual work of these men, as recorded in the labor sheets, indicated that a much larger reduction could be made, and that with the electrical engines out of operation, it would be quite possible to so reduce the hours of day and night labor as to operate the staff with a complement of eight men in the winter service and five men in the summer service, which would have had the effect of reducing the labor cost by $8,760 per annum, a reduction of more than 45 per cent.

This was largely due to the unquestioned fact that the amount of fuel required to be handled would be greatly reduced, and that there would no longer be the least excuse for the maintenance of the operation of machinery at night during all times in the year, and that during the summer season, by utilizing electricity for the elevators and house pump services, there would be no reason for operating steam boilers at all.

By the purchase of both steam service and electrical service, the labor conditions could be still further modified, and it was suggested that if this course were followed the force could be reduced to four engineers. These views were, however, resisted very strenuously by the representatives of the Department, who were admittedly affected in their views by the influence of labor organization representatives, who were constantly on hand during the trial, and kept themselves familiar with the conditions obtaining there, and were careful to let their views be known to the officials.

The question was practically solved by the proposal of the New York Service Company, an organization which is engaged in maintaining a number of steam-power and heating plants in this city, who upon a careful investigation, offered to undertake the operation of the steam and pumping plant, in combination with a purchased supply of electricity, for figures which were substantially less than those disclosed by the test for any given amount of steam which might be found to be required under the changed conditions. This offer included the employment of labor of the city employes to the necessary extent, upon the same rates of wages and the same hours of labor as under direct municipal management.

The entire expense of the operation of the plant for the year of trial was $62,229, inclusive of all elements, and the offer of the New York Service Company, together with the cost of electric energy as disclosed by the trial, in the quantity necessary to maintain all the services of equal extent, amounted to a total of $50,533. This was upon an estimated amount of steam which the representatives of the New York Edison Company considered would amply supply the equivalent of the services developed during the year 1913.

An alternative offer was made upon another estimate of steam requirements, made by representatives of the Bureau of Municipal Research, the total cost being $51,640.

A third offer was made, based upon the value

of liberal estimates of requirements for steam made by the representatives of the City Department, at a price of $53,326.

It will be observed that all these proposals were substantially less than the total cost of operation of the plant during the period of the trial.

Accountants who made an independent investigation of the figures determined by the test, also demonstrated that if new buildings of the same character, and with the same requirements, had been constructed, and no power plant had been installed, the steam plant could have been operated for the heating of the building, and electric energy could have been purchased for exactly equivalent services, at a total annual cost of from $43,400 to $46,200. These figures confirm the general conclusion that the installation of this plant and its operation, even under the most careful and economical municipal management, involved the city in a continuing loss of the difference between its existing total expense and that which would have obtained if no such power plant had been installed.

CHAPTER VII.

THE COST OF PRODUCING STEAM.

THE production of steam under pressure is the process which underlies all the operations of such power plants as that in the Hall of Records building. It involves the employment of skilled labor in the firing of coal, the purchase, storage and handling of the fuel, the removal of ashes and clinker formed in the process of combustion, the supply and heating of the water which is to be evaporated into steam, and finally, the lighting and ventilating of the space occupied by the firing staff. There are in addition to the foregoing, expenses in the upkeep of the boilers, the repair of worn or broken parts, the renewal at regular intervals of various parts of the furnaces, especially of the grates, and the inspection and cleaning of the exterior and interior of the boilers, much of which work must be done by skilled mechanics, whose services are an addition to the expense of the firing labor.

In the Hall of Records the combustion of the fuel also involved the operation of a forced draft blowing fan, operated by a steam-driven engine, the expense of operating which naturally became a charge upon the cost of the steam which was produced.

During the trial a special effort was made by the Bureau of Public Buildings to secure an efficient operation of this part of the plant, with fairly good results in the effective combustion of the fuel. This fuel varied in character during the test. For the first five months, a mixture of bituminous or "soft" coal and anthracite or "hard" coal, was utilized at an average cost of $2.86 per ton of 2,000 pounds. The use of this mixture, even with the most careful methods of control of the air supply, resulted so frequently in the emission of volumes of black smoke from the chimney, that the attention of the public press was drawn to it and some articles appeared in the papers, reflecting upon the Department in charge of the building, as being violators of the regulations of the Department of Health in this respect.

As a result of this situation the Bureau of Public Buildings abandoned the use of the soft and hard coal mixture, and for the seven succeeding months used anthracite fuel of the size and quality known in trade as No. 1 Buckwheat, the average price of which was $3.30 per ton.

Throughout the entire period of the test the average cost of fuel thus became $2.95 per ton, and the total amount which was burned was 5,166 tons, by means of the combustion of which a total of 81,653,979 pounds weight of water was evaporated into steam.

The relation of the fuel to the product was of a fairly satisfactory character, being 7.9 pounds of water evaporated per pound of fuel burned. This result was partly secured by a method employed by the Bureau, which stimulated the interest of the firemen in aiding the demonstration of efficient operation of the plant. A large blackboard was hung in the fire-room, upon which was written at intervals the result of the work of the firemen during each watch. This induced the men to effect the best results during the period of their watch.

A part of the observations of the trial extended

to the analysis of the gases discharged from the boiler furnace to the chimney. A very elaborate apparatus was provided for this purpose, and careful note was taken of the efficiency of the combined process of burning the fuel and transmitting its heat to the water.

The average result during the entire period of the test was an efficiency of 69 7/10 per cent. This may be regarded as a good result under commercial conditions.

As has already been mentioned, the effort made by the Department towards efficient operation of this plant resulted in a reduction of the labor which had been previously employed in the fireroom, the number of men being reduced from eleven to eight. This very plainly indicated the extravagant extent of the employment of labor under normal municipal operating conditions, the reduction of the number to those who were found to be sufficient for the work done during the trial, being 37 per cent.

Even under these conditions the cost of labor in relation to the fuel handled was excessive in comparison with equivalent conditions under private management, due in great part to the municipal system of eight-hour service, involving three watches of the firemen per twenty-four hours. The result of this arrangement was a total cost for labor per ton of fuel handled of $1.67, or an addition to the average value of the fuel of more than 50 per cent.

The employment of an ample staff of firemen upon a three-watch basis, results in much irregularity in the extent of actual labor exerted by the men. The total number employed is naturally that which will be required to handle the maximum amount of fuel expected to be consumed at some particular time. The result is that these men have very little to do at times of minimum demand for steam. The observations during this prolonged test showed that a maximum condition existed only during one week of the year, the eighth of the test, the period being that of the very coldest weather. At this time and under these conditions, the fuel required to be handled was very easily within the capacity of the men who were employed.

It happened that shortly after the conclusion

of the trial a very unusual combination of extremely cold weather with a strong wind arose on the 14th of January, 1914, bringing about a demand for the consumption of fuel which was most unusual in extent. Observations on that day in the Hall of Records showed a consumption of 35 tons of fuel, which was more than double that which was consumed during the day of greatest demand during the period of the test. This unusual quantity, however, was handled without difficulty by the same staff that had been employed during the trial.

It will be evident from these observations that the cost of the same labor when employed upon the combustion of the much lessened amount of fuel in summer time would very greatly enhance the cost of steam production during that period, and it was found that the cost of labor per ton of fuel consumed during the summer rose to $2.40, practically doubling the cost of the fuel.

Under such circumstances, which will be found to be existing to some extent in all plants in business buildings, it becomes a question whether the cost of the steam in such reduced quantity is not in excess of that at which it could be purchased from a public supply, even at a relatively high price or rate. So pronounced is the cost of the production of the small amount of steam often required in small buildings, that it even becomes probable that the use of gas at prevailing prices may be a source of economy. In particular, during many hours of summer night service in the Hall of Records no steam at all was raised, and the watch of firemen employed during that time were performing no productive service.

One of the special objects of this trial was the comparison of the cost of the production of steam with that of a supply purchased from the service of the New York Steam Company, the supply provided by which Company would have avoided the entire expense of labor and fuel, and would also have rendered unnecessary the use of the steam blower, the feed-pumps, return-pumps, blow-off or drip-pumps, the feed-heater, and the use of the electric ventilating fans and the gas and electricity for lighting the fire-room. It would also have rendered unnecessary the cost of the removal of ashes and the use of city water.

Further, had this service been adopted prior to the installation of the boilers and their appurtenances, it would have saved the entire investment in boilers, the cost of which was approximately $32,850, and the fixed charges on which amount to $3,056 per annum.

Further, the amount of steam thus required to be purchased would always have been less than that which was required to be produced in the boilers, because no steam would then have been required to operate the auxiliary pumps, which alone consumed 10 7/10 per cent. of all the steam which was generated during the trial. It would thus have come about that the amount of steam required to be purchased would only have been 87 per cent. of that which was necessarily produced in the plant. It would have been quite possible to utilize the New York Steam Company's supply to continue the operation of electric and hydraulic engines and pumps, or it could have been purchased for the purpose of the heating of the buildings if those engines and pumps should be discontinued.

It would also have been possible to arrange for

the supply of this steam during the summer season, continuing the operation of the steam boilers during the winter or heating season. This arrangement, however, presented the difficulty of dispensing with the firemen's labor during the summer season, and their re-engagement for the winter season. This is a process which can be, and frequently is done in private operations, as many of the firemen find employment during the summer season on steamboats, and at its conclusion are available for employment in the buildings and power plants. But municipal employees cannot so readily be discharged and re-engaged.

It must be noted that the price at which the New York Steam Company would sell steam to the city, which at the time of the trial would have been about 42c per 1,000 pounds weight of water in the form of steam, would be accompanied by an advantage to the city of the payment by the Company for the water which is thus used and turned into steam. The New York Steam Company has been paying for water used in this manner for many years about $40,000 annually, at the regular price of $1 for each 1,000 cubic feet of water.

Thus in purchasing its steam supply from the New York Steam Company, the city would be securing an additional revenue from that Company by payment for the water used for its supply.

It may be noted that the Steam Company naturally uses more water per 1,000 pounds of steam sold than the actual weight of the amount thus delivered, since it has various losses by condensation in its plant and piping, which have to be provided bv water purchased from the city. But even taking the net amount in question, it will be found that the city gains 4 per cent. upon the cost of the steam by this return in the form of payment for water.

The expense of the operation of steam production involved that of the removal of ashes from this plant, the process proving during the period of the trial to be one in which the element of humor was somewhat conspicuous. It was alleged by the Bureau in charge of the operations that the disposal of these ashes could be effected without any expense to the city, and that in point of fact, they might even be made a source of profit by being sold for building purposes, and that in any case they would be removed by wagons hired and utilized by the Department of Public Works at less expense to the city than by the usual means of the wagons of the Department of Street Cleaning.

The latter was the method of their removal prior to the trial. That Department conducts this operation for all city departments, as well as for the general public, at a cost which averages 74 cents a wagon load, or about 5 cents per 100 pounds weight. The quantities have become so large and the expense so heavy, that the Department has of late years refused to undertake the removal of the ashes from large steam plants, and has required their owners to make arrangements with private contractors for the purpose.

The prevailing price paid for this work is the same figure, 5 cents per 100 pounds.

For the purpose of economical utilization of the ashes the Department of Public Works hired from a contractor a team and driver at a cost of $3.50 per diem, and started a process of removal of the ashes which they claimed were worth at least $5 a day for filling-in purposes of the Bureau of Highways.

A little observation of the circumstances that followed readily indicated that the ashes were being disposed of in a very different manner, and without informing the officials or the teamster in charge, the movements of the latter were followed by observers, whose reports became most amusing in their disclosure of the methods that were followed. These observations showed that approximately one-third of all the loads of ashes were taken directly to the scows of the Department of Street Cleaning and were thus disposed of at public expense, that approximately another third were dumped on vacant city properties under conditions adverse to their appearance, if not to the public health, and only about one-fourth was usefully utilized in some municipal building operations then in progress.

Destination of Ashes	No. of Loads or Portions of Loads Carted Away	Per Cent
Used at the New Municipal Building	21	25.3
Strewn under Manhattan Bridge.....	17	20.5
Strewn at rear of Children's Court Building	10	12.1

Scattered in Street	9	10.8
Carried to the East River Ash Dump of the Dept. of Street Cleaning . . .	26	31.3
	83	100.0

The methods under which the removal was effected by the municipal employees were illustrated by the following observations, being one of a number made from time to time over a period of seven months of the trial.

May 6, 1918

"We reported at Reade street side of Hall of Records at 7:50 a.m. on the 5th instant, and found single cart marked 'Bureau of Highways, Manhattan, No. 80,' loaded with ashes; left at 8:02 a.m., went to Clinton street dump, arriving at 8:23 a.m., dumped same on scow; came out at 8:33 a.m., went to Kelly & Lowery saloon, 259 South street, where driver entered at 8:40 a.m., came out at 8:45 a.m.; driver entered saloon, 51 New Chambers street at 8:55 a.m., out 8:58 a.m.; returned to Hall of Records at 9:05 a.m.; left at 9:40 a.m., went to Clinton street dump at 10:03 a.m.; came out at 10:11 a.m., went into saloon 259 South street 10:20 a.m.; came out at 10:24 a.m.; went into harness store, 68 New Chambers street, at 10:28 a.m., out at 10:30 a.m.; returned to Hall of Records 10:35 a.m., left at 11:00 a.m. and met another driver; they both entered saloon 199 South street at 11:19 a.m., out at 11:24 a.m., Clinton Street dump at 11:28 a.m.; came out at 11:40 a.m., drove to pier 35 East River, where he stopped and fed horse at 11:45 a.m.; left at 12:40 p.m., returned to Hall of Records at 1 p.m.; left at 2:03 p.m., went into saloon, 51 New Chambers street, at 2:14 p.m.; came out at 2:17 p.m.; Clinton street dump at 2:35 p.m.; came out at 2:42 p.m.; left cart in

Ash Disposal by Municipal Methods.

front of dump and rushed the can with driver of cart marked 'Dept. of Ferries, No. 12'; left at 3:20 p.m., went into saloon 259 South street 3:30 p.m.; out at 3:36 p.m.; went into saloon 59 Stanton street at 3:54 p.m., out at 3:59 p.m.; went into saloon 598 Second avenue at 4:40 p.m. with another man, came out at 5:00 p.m. and went to stable at 320 East 35th street, arriving at 5:10 p.m."

These investigations were finally brought to the attention of the Department and led the Department to admit that the cost of the removal of ashes had been found to be as much as would have been the case had they been removed by the Department of Street Cleaning at the regular rate of cost.

The cost of removal during the year of trial was found to have been $1,219.98, or about 23 6/10 cents per ton of coal.

This may be compared with the average cost of the removal of ashes in similar power plants under private management, which would be 14 to 17 cents per ton of coal.

The average proportion of the weight of ashes, which was ascertained during the year of trial, was 18 4/10 per cent. of the fuel, which indicates a fair quality of coal. The cost of removal by the methods employed was, therefore, about 40 per cent. in excess of the commercial cost.

The removal of these ashes, which amounted to a total of 949 tons during the year, involved in this building, as in many others, the use of an hydraulic elevator. In the Hall of Records the floor of the boiler-room is twenty-five feet below the level of the street, and an hydraulic plunger elevator is provided for the purpose of lifting ashes up to the sidewalk and returning the empty cans to the boiler room. A record of the operation of this ash elevator was made during the year of trial, and it was found that it ran a total distance of 188½ miles. Its operation formed a part of the work of the elevator power plant, and its cost could thus be reasonably ascertained, the value of the steam used for its operation being $41, and the share of the total expense of elevator operation about $117, or 2¼ cents per ton of coal used in the plant.

In order to ascertain the full cost of the removal of the ashes, there should also be considered the cost of the laborers' time in handling the cans and loading the ashes in the wagon. About one-half of the time of a laborer in the fire-room was thus occupied, at an expense for the year of the trial of $460.

These figures thus afforded a total cost of $1,797 as the expense of getting rid of the ashes from the fire-room, or per ton weight of ashes, $1.89, or per ton weight of fuel burned, 34 8/10 cents.

A minor element of expense in this as in other plants is the use of gas in the boiler room, and in the rooms containing the pumping and auxiliary machinery, as a supplementary supply of light in case of the failure of the electric service. The maintenance of light under every possible contingency in a generating plant where high-pressure steam is employed, is of vital importance. In this plant it was found that the consumption of gas during the year amounted to 341,200 cubic feet at a net cost to the city of $252.30.

With the foregoing information in hand, it became possible to determine the total cost of the production of steam by the fuel and labor thus employed, and to compare the results with the purchasable price of an equal supply.

The cost of production was as follows, during

the test period of the year 1913:

Fuel	$15,296.58
Labor in the fire-room	8,641.25
Removal of ashes	1,219.98
Gas	252.80
Supplies, repair work, and the repair of the boiler and auxiliary apparatus, by Department records, was	982.25
Supervision	1,680.62
Total	**$28,022.98**

Representing an expense per 1000 pounds of water evaporated into steam, of 34 cents.

In addition, had no boiler plant been originally installed there would have been saved the annual expense for interest and depreciation upon the investment in these boilers, amounting to approximately $5,676; and the space that they occupied could have been realized for other purposes, at a rental value of $3,684.00, a total of $37,-382.98.

Turning now to the cost of equivalent service from the public supply, it would be found that, owing to the lessened demand for steam by reason of the elimination of the auxiliary machinery, it would only have been necessary to buy from the New York Steam Company a total of 70,-972,879 pounds weight of steam.

This at the price of 42 cents per 1000 pounds, would have cost	$29,808.00
This supply would have brought back to the city in the form of payment for the water used by the Steam Company, a sum of..........	1,120.00
The net expense for this steam, therefore, would have been	28,688.00
To which should be added the cost of supervision at the same rate as that required in connection with the plant, or	1,824.00
The total cost of this purchased steam supply, therefore, would have been	30,512.00

Against this the cost to the city of its own plant's operation with supervision, interest, depreciation, and the rental value or annual expense of the space occupied by the boiler plant was **$37,383.00**, representing an annual loss of **$6,871.00**.

This makes it perfectly evident, therefore, that the City's investment in this boiler plant has no commercial value, since it effects nothing that could not have been equally well done at the same, or less cost, by the purchase of the supply without the necessity of the investment of the taxpayers' capital. Moreover, it may be noted that, as later observations will indicate, several other available methods of operation in combination with the public service of electricity, would have reduced the amount of steam required to be taken from the New York Steam Company to a very substantial extent, and it would have been possible to purchase no more from that service than that required to effect the operation of house heating and the warming of hot water.

This situation disclosed by the accurate figures secured during this prolonged test, is a conclusive demonstration of the undesirability of the investment of municipal funds in machinery, the purchase of which can be equally effectively performed by the utilization of the services of public corporations, where these services are available.

There are indirect advantages, of course, which increase the desirability of the purchase of such services from an enfranchised corporation, inasmuch as the city thereby gains whatever increased amount of taxation or franchise returns accrue from the increased business placed with the corporation.

By engaging upon the operation of production of steam as well as of electric energy the municipality embarks in competition with its own partner, the corporation which it has authorized to conduct the same line of business for the general convenience of its citizens. A community thus discounts its own interest in the utilization of its franchise property, to the extent to which it invests in machinery that reduces the business of the public system.

CHAPTER VIII.

THE HEATING OF THE BUILDINGS.

THE interior of the Hall of Records and the three exterior buildings is warmed by the use of radiators which are supplied with steam through piping connected to a common point of supply in the Hall of Records. The pipes between the several buildings are laid under-ground in trenches and consist of a supply-pipe large enough to allow of the flow of the exhausted steam, and a return-pipe which contains the water or condensed steam after it has passed through the radiators and is thus conveyed to the Hall of Records and pumped back into the boilers.

The operation of conveying the steam to radiators situated at a considerable distance from the point of supply, involves some pressure in the steam, usually from five to ten pounds per square inch. This pressure is required to overcome the friction developed in the passage of the steam through long lengths of piping and around a number of bends and turns, and it is also necessary in order to expel from the radiators the air which collects inside those appliances during the process of condensation of the steam. The air, which is always more or less present in steam, finds its way into the steam boilers with the feed-water, all water being charged with some quantity of air in the form of minute bubbles.

This air has to be forced out from the radiator, usually through a little device known as an air-valve, which is placed on the end of the radiator opposite the point of supply of steam. This operates by opening a very small orifice when air is collected near the valve, but closes the orifice as soon as steam reaches it. Usually the air thus expelled is driven into the room in which the radiator is placed, but where exhausted steam is used the air is often quite foul-smelling, due to the presence of some oil in the steam, so piping is provided by which the air is led to some point where it may be discharged without causing annoyance.

It is possible to operate these devices by the use of steam at the moderate pressure mentioned without any mechanical devices other than the air-valve described, but so soon as an attempt is made to operate at less pressures such as are necessary when exhausted steam is used, then some special apparatus must be provided whereby the air may be withdrawn from these radiators by mechanical means, as the steam will not have sufficient pressure to force it out. When therefore, it is desired to utilize exhausted steam from engines or pumps, this supply must either be maintained at a pressure in excess of five pounds, which would be detrimental to the economical operation of the engine or pump, or special arrangements must be made for the purpose above described. The use of exhausted steam therefore, for the purpose of such heating work is accompanied by the necessity to provide and operate air-pumps or air-ejectors which draw out this foul air, and in certain cases also draw away the water from the radiators. It will be seen therefore, that the operation of these pumps is an additional expense involved in the use of exhausted steam and reducing its value to that extent.

The use of exhausted steam from appliances such as pumps and engines, would be as fully economical as could be expected if it were possible to so proportion the operation of the engines or pumps and the amount of steam which they use at all times, as to fit in with the requirements for a supply of steam for heating purposes, but this unfortunately can never be wholly accomplished. Climatic conditions in the first place, vary by locality and by all the numerous gradations in temperature which are found in different parts of the country, and it will be evident that the demand for heating steam would be in greatest volume and extent and for the longest period in the northern sections of the country, while there would be little or no demand for such purposes in the southern states.

In the latitude of New York the situation is about half way between these extremes, and the variations in temperature range during the heating season from 70 degrees, at which no heat is required, down to a rare and limited occurrence of zero temperature.

The average temperature during the entire heating season, which extends over a normal period of about two hundred days, is 38 degrees, and therefore, the average increase in the temperature inside the building over that prevailing outside, is the difference between 38 degrees and 70 degrees. But this statement is a mere broad view of the average conditions, and by no means conveys an impression of the extraordinary variations which in this climate occur from hour to hour, between day and night, and from day to day during the season of cold weather.

It will readily be seen therefore, that the use of the wasted steam exhausted from engines or pumps can only be expected to be proportioned to meet some one or other of the numerous combinations which are thus brought about in the demand for heating steam. Thus the greatest requirement for the work of the electrical engines in a public or business building occurs in the early evening hours, and yet the coldest part of the day is usually the early hours of the morning. If elevators are operated the steam given out by their pumps is fairly constant throughout the hours of the business day, whereas the demand for steam may vary from a requirement for the whole of such steam, or even more, down to nothing at all.

There are some small appliances in connection with a power plant which are in quite constant operation, such as the feed-pumps and small auxiliary apparatus which give out steam often in somewhat wasteful proportions, at all times of the day and night, while the plant is maintained in operation. This limited supply of exhausted steam however, can usually be utilized best in another service of an economical nature, namely, the warming-up of the water which is fed into the boilers. It is rarely in sufficient quantity to afford any surplus available for the purposes of house heating. The combination therefore, which was chiefly under discussion in connection with this test, is that which contemplated the utilization of the exhausted steam of the elevator and house pumps, and especially that of the electrical generating engines.

As regards the quantity of steam which is required in a heating service in the climate of New York City, a number of observations prior to this test had established the fact that the average use during the five thousand hours of the heating season was 4/10 of that which would exist upon the very coldest day. In other words, supposing that the heating system on a zero day in February should condense 1000 pounds of steam, it would follow that on days when the temperature was higher it would use less than that amount, and as the temperature rose near 70 degrees, the demand would cease altogether, the result of the operation being that throughout the entire period of heating the average quantity used per day would be 400 pounds.

The test in the Hall of Records, conducted with the utmost exactitude by measurements from hour to hour over the whole heating season, confirmed this conclusion. When the plant was originally installed in the Hall of Records Building, it was arranged that the exhausted steam from the engines and pumps should be supplied to its heating radiators. This afforded a situation in which the exhausted steam during the working hours of the day was substantially in excess of the heating requirements, excepting in very extreme weather, when it was necessary to add to the exhaust a certain amount of steam taken direct from the boilers, commonly referred to as "live steam," which is reduced in pressure to the same pressure as the exhaust by its passage through what is known as a reducing valve.

The proportions of the four buildings are about 10 million cubic feet, of which the Hall of Records was about 45½ per cent., the County Court, 31 per cent., the City Hall, 15-3/10 per cent., and the City Court, 8-2/10 per cent. The use of steam did not follow these proportions, because the Hall of Records was found to use 60 per cent. of all the heating steam due to the effect of its ventilating system. The other buildings may be assumed to have used the remainder of 40 per cent. of the total, in proportion to their dimension, or 57 per cent. in the County Court, 28 per cent. in the City Hall and 15 per cent. in the Municipal Court.

The addition to the Hall of Records services of the work of heating the three exterior buildings, increased the possibilities of the use of the exhausted steam by 175 per cent., and thus rendered it a much more economical combination, since it then became possible to utilize to a greater extent during the milder weather the wasted steam of the engines and pumps. But the difficulty of interrelating the heating apparatus to these machines and the extent of their work was found to preclude a greater utilization of the steam which they afforded than about 54 per cent. Out of the production of steam during the whole year, only about 35 per cent. of the entire quantity produced was utilizable in the form of exhausted steam for heating purposes. It had been claimed by advocates of the combination of exhaust-steam heating with electric generation, that at least as much as one half if not more of the total steam supplied to the engines could be thus utilized.

The amount of exhausted steam in relation to the electrical output will of course, vary with the type of steam engine employed. The more economical the engine the less quantity of steam will be used for any given amount of work, and therefore, the employment of a wasteful engine will give more steam in the form of exhaust. The more wasteful is the engine, the more apparent gain there is in the form of the use of exhausted steam. The process is similar to that of wasting at the bung and saving at the spigot.

The engines in the Hall of Records building are fairly high class, economical engines, of what is known as the Corliss type, and used a fairly average rate of steam in proportion to their work. They consumed about 48½ pounds weight of water per hour in the form of steam for each kilowatt of electric energy they generated. On account of the fluctuation in the demand for electricity it was necessary to operate a smaller engine of a more wasteful character, which during the time of small loads, as at night, operated with a fairly constant load and near its capacity. This little engine used 75½ pounds weight of steam per k.w. hour, and it was the source of much of the exhausted steam that was utilized.

The general result of this system of operation was that the average consumption became 50.9 pounds weight of steam per kilowatt hour. So the steam which was available for the purpose of heating was procured not only from the economical engines, but from the much more wasteful type of engine.

The small engine used 75.5 lbs. per kilowatt

hour as compared with 48.5 lbs. used by the Corliss engines, or 27 lbs. per k. w. hr. more than the latter. It produced 62,665 k. w. hours out of 712,763 or about 9 per cent. of the total electricity, using 13 per cent. of the steam.

The contribution by the wasteful engine to the work of heating was not wholly economical. It represented steam that had been used in a wasteful way of generating electricity, which thus created additional exhaust steam available for heating. This had no more economic effect than would have been effected by passing that amount of steam direct from the boilers to the radiators.

In considering the proportion which the various appliances provided waste or exhaust steam for heating purposes the Board of Engineers concluded that it would be proper to regard the exhaust provided by the house, sewage, and elevator pumps as being that which was the main source of supply, inasmuch as these pumps would be continued in operation should the electrical plant be dispensed with, and the exhaust derived from the electrical engines was regarded as an addition to that available from the pumps.

It was found that the contribution to the work of heating provided by these electric engines was 32 3/10 per cent. of the total amount of steam which was supplied to those engines during the two hundred and twenty days of the heating season. This result was one of the answers to the main subjects involved in the discussion which had led to the determination to conduct this test, and it very clearly demonstrated the exaggerated nature of assertions as to the results to be secured by the use of exhausted steam, made by those who had not fully studied the subject in all its details. Such assertions have sometimes assumed that the electricity generated in power plants in which the exhausted steam was used in the work of heating, was produced without any cost, but inversely, those who are more interested in economies of steam heating than in those of electric generation, assumed that the heating work had been produced under such circumstances for nothing. Both views are, of course, fallacious in that neither took into account the limited extent to which the combination can be made to coincide in production and use.

Part of the work of heating in which a minor use was made of exhausted steam was the heating of the warm water required for sanitary purposes. This was not a very large amount, requiring during the entire year, only 1,548,050 pounds weight of steam, or out of the total exhaust utilized in heating work, only about six per cent.

It may be noted in this connection that the use of hot water in these public buildings was extended throughout the entire year, and it would therefore, have appeared possible to carry on this small service entirely by the use of exhausted steam. But even this small demand could not be fitted in with the supply of exhausted steam available at all times, and it was necessary to add from time to time seventeen per cent. of its total in the form of live steam.

It was observed that the greatest demand for hot water did not occur in the period of coldest weather, but in the week ending April 20th, when the demand reached 65,785 pounds. The average weekly consumption during the year of the test was 36,200 pounds, and in the warmest summer weather there was still required as much as

18,192 pounds in a single week.

The entire demand for the work of heating and hot water involved the use of 32,317,287 pounds weight of steam, of which 27,497,764 pounds was provided in the form of exhausted steam, or nearly eighty-five per cent., the difference being provided by steam taken direct from the boiler. Of these totals the heating of the Hall of Records building involved only 11,136,638 pounds, or less than thirty-five per cent., and it is interesting to note that the work of heating this building required only 13 6/10 per cent. of the entire quantity of steam raised during the trial.

The largest demand in the building for a single week of severely cold weather was 716,000 pounds weight of steam, which could have been provided by any one of the five boilers installed in the building. The largest requirement for the heating of all the four buildings was found to be 233,000 pounds of steam in one day, which was approximately equal to 310 boiler horsepower, and was well within the capacity of any two of the five boilers installed in the building. It is evident therefore, that the larger part of the investment involved in this large steam-generating plant, both as regards its space and apparatus, was attributable to the addition of the power-generating plant and was not required for heating purposes. Moreover, the heating of these buildings, if unaccompanied by power apparatus, could have been effected by low-pressure boilers of a much cheaper kind of construction than those that are installed.

Reference was made in the earlier part of this description of the heating of the buildings to the additional apparatus involved in operating the air-systems of these radiators. The three exterior buildings were provided with apparatus known as the Webster Vacuum System, by which a suction is maintained upon the outlet-valves of each of the radiators, through which the condensed water, and also the air that collects in the radiators, is drawn. The suction is produced by steam-pumps in constant operation, and these pumps used during the heating season, 2,451,847 pounds weight of steam, which was 13 per cent. additional to the steam used in heating these buildings. It is true that the steam thus used was in part exhausted back into the pipes from which the heating steam was drawn, but inasmuch as it thereby displaced the use of exhausted steam to some extent from the main engines, it reduced the amount to be credited to the operation of the electrical plant, of which it practically formed a working part.

The circulating system in the Hall of Records building was of a different nature. It is that known as the Paul system, by which a suction is maintained upon each of the air-valves of the radiators connected through a system of piping to steam jets blowing high-pressure steam through ejectors, which steam is of course, entirely wasted. This apparatus used 275,678 pounds weight of steam, or 2½ per cent. additional to the steam used in heating that building.

These circulating systems are, therefore, to be regarded as forming a part of the necessary additions to power plants, being accompaniments of the use of exhaust-steam heating, which are not required if the same results were to be effected by a direct steam supply.

It would be unnecessary to use either of these

systems should the power engines and pumps be shut down and the use of exhausted steam be abandoned. It is, therefore, proper in considering any distribution of the costs of operation that the expense accompanying these circulation systems should be added to that of the operation of the generating and pumping apparatus and should not be charged to the cost of the work of heating, the commercial value of which is clearly no more than would be the cost of purchasing the steam for equal heating work, or of a direct process of raising low-pressure steam sufficient to circulate throughout the system, without the use of these appliances.

These considerations and facts brought together the figures of the cost of the work of heating the buildings and warming the water for sanitary purposes, which was found to be as follows:

Steam, 40.57 per cent of all produced	$11,365.59
Engine-room labor, supplies, repairs	1,576.23
Cost of space, interest and depreciation	3,795.00
Total	$16,736.82

The average cost per 1,000 pounds being thus, 50 4/10 cents.

The determination of the volume of steam required for this service made possible a definite comparison of the cost of operating the same heating systems by the use of steam purchased from the New York Steam Company's service, which service would eliminate the pumps and steam jets, and would thus reduce the requirements to a quantity which at 42 cents per 1000 lbs. would have cost only $13,180.00.

The return to the city paid by the New York Steam Company for water would then have been $520, leaving the net cost at.....	$12,660.00
To this should be added supervision at the established rate, amounting to	805.00
The total cost of the New York Steam Company's service would therefore have been only	13,465.00
Or only $520 more than the actual expense of the production of the steam used during the year of trial, and cheaper than its total cost by	3,271.00

According to the point of view of the subject, the exhausted steam was either a gift or a large by-product of the power plant to the work of heating, or inversely, it was an unavoidable expense at its full value to the work of heating, and the other services became partly by-products of its use in that work.

If the exhaust steam had not been utilized and debited to the cost of steam heating and thereby credited to the cost of each of the other five services, the effect would have been to charge back to the machinery $10,025 as the cost of the exhausted steam, which was provided out of the machinery in the following proportions:

Service	Steam, lbs.	Per Cent of Total	Cost Divided	Adding to Their Cost
To the electric engines	11,118,656	53	$5,300	.74 of a cent per k.w.hr.
To the elevator pumps	5,207,772	24	2,400	15 cents per car mile
To the house pump	3,137,552	15	1,500	50 cents per 1000 cu. ft.
To the refrigerating pumps	890,306	4.4	440	16 per cent on the cost
To the sewage pumps	735,624	3.6	360	30 per cent on the cost

This shows the limit of the value of the utilization of exhausted steam, even upon such very unusually advantageous circumstances as the increased heating surface of three buildings being added to the original opportunity for use of such steam.

Surplus Exhaust in Coldest Weather.

On the other hand, if the view should prevail that the exhaust is a gift to the heating system, then it reduced the cost of that service by $10,025. out of $16,736, or 60 per cent. But either view is incorrect. The fact is that the exhausted steam had no more value than the price at which it could have been bought, which was less than $8,800, and it formed a part of a combination, each element in which is relatively expensive, and which could be reduced in expense as a whole by a change to other methods. Under such a rearrangement it might even pay to purchase or produce steam for heating at a relatively high cost, provided that economies could be effected in other directions resulting in a total less than that which was found to prevail.

CHAPTER IX.

THE COST OF ELECTRIC ENERGY.

IN the subject of the electrical energy generated and used we reach that part of the trial which was of most particular interest. The electrical service was that on which centered the discussions that led to the establishment of the test, and the determination of the real character and the true cost of this particular part of the work of the plant was perhaps the most important feature of the observations.

The use that was made of electric energy produced by the plant was noted hour by hour on recording instruments, which determined the exact amount utilized in the requirements of the several buildings. An independent record was maintained of the amount of energy used in operating ventilating appliances for cooling the engine and boiler-room space, also for the lighting of the spaces occupied by the machinery, and the energy involved in the operation of the storage battery.

As an illustration of the self-imposed burdens of an isolated plant, it was found to be consuming 14 per cent. of all the energy which it created for the purpose of lighting and ventilating the space in which the work was carried on.

Most of this energy formed no useful part of the work accomplished by the machinery, inasmuch as the larger part would not have been required if the plant had not been in operation. The quantity of electricity generated during the year of the trial amounted to 712,763 kilowatt hours, of which amount 136,310 kilowatt hours represented energy diverted to the purposes of service of the plant. A subdivision of this amount showed that the mere lighting of the spaces occupied by the machinery absorbed 39,777 kilowatt hours, or 5½ per cent., and the ventilating fans 84,266 kilowatt hours, or nearly 12 per cent., while the losses due to the operation of the storage battery amounted to 12,267 kilowatt hours. The net amount of energy which was distributed for the service of the buildings in the form of light and electrical power, was only 576,453 kilowatt hours, or about 80 per cent. of the total that was created.

It must be recalled here that the original purposes of this plant for which it was planned and proportioned, was the service of the Hall of Records building alone, and this was found to be but seven-tenths of the entire energy created during the trial. If the plant had been thus limited in its service, the proportion which the maintenance of light and ventilation in the sub-basement would have borne to the total would have been nearly thirty per cent. The information thus secured was very illustrative of the burdens which are supported in the operation of power plants for such restricted purposes as the service of a single building. It also demonstrated the usefully-employed amount of electricity which would have been required to be purchased for equivalent service, if the supply had been derived from a public system.

A study of the circumstances resulted in the determination of these net requirements for the Hall of Records and the other buildings, which it was found would be covered by about 84 per cent. of the total which had been generated, or approximately 593,763 kilowatt hours. This amount made a fair allowance for necessary lights in and around other parts of the machinery spaces than that in which the electric engines were situated, and also provided for part of the energy being used in ventilating the boiler-room space should the boilers be continued in operation. By this means definite information became available upon a part of the original question at issue, namely: what was the actual value of the energy generated per kilowatt hour, on which basis the electricity would be purchasable from the public service.

It will be readily seen that if the entire amount which had been created should be divided into the entire total cost, the result would be a low price per kilowatt hour, or as it is technically described, "per unit," whereas, if the net amount which would be required to be purchased be regarded as the useful product of the plant, and that amount be divided into the total cost, it would show that the cost was a higher price per unit. Such a higher figure is, of course, favorable to the alternative system of the purchase of the electrical supply. In many cases it will be found that those who are operating such power plants are basing their ideas of the apparent low cost at which the energy is produced, by assuming that all of the product is of a necessary character, and is usefully utilized, which cannot in any case be the true situation.

There is in addition to these considerations, another feature of considerable importance in the case of a municipal plant producing electric energy in competition with the supply of a public company which is paying taxes to the city. In such a case the municipal operation is competing with the output of its partner and reducing the total business performed under a franchise, thereby reducing the tax-paying capacity of the public service corporation.

The payments of taxes over a period of years by the New York Edison Company to the City of New York were investigated and it was found that the taxes paid bore a proportion to the amount of energy sold by that company sometimes as high as 3/10 of a cent per kilowatt hour.

The Committee of Engineers agreed that the relation which existed between the output and the taxation paid was not less than 2/10 of a cent for each kilowatt hour of energy that had been sold, and this figure was adopted by the Board of Engineers as representing the minimum return that would have been made by the New York Edison Company to the City, in respect of the purchase from that Company of the energy required for these buildings if the plant had not been in operation. It could most easily be understood and could most properly be regarded as a credit to the purchase, which is equivalent to a reduction of the price of purchased energy. Upon the amount which was found to be required it represented an annual return or reduction in cost to the city amounting to $1,187, or a reduction of the price per kilowatt hour to 1.92 cent.

At the time the trial was undertaken, the contract then in operation in regard to the group of City buildings was based upon the lowest prevailing price for energy, or three cents per kilowatt hour. In order to reach the quantity which justified this price under the contract, an agreement had been made to include in the bills for these City buildings the quantity of electricity used in lighting the Brooklyn Bridge and its approaches.

When the trial was undertaken and the City buildings were supplied from the Hall of Records power plant, it left the service of the Brooklyn Bridge as the only one of the group supplied by the public service, and this necessarily involved a higher rate being paid during the year of the trial for that service, which was found to have amounted during the year of trial to $279.34. This, of course, added to the real cost to the municipality of the operation of the plant.

During the period of the trial the methods under which the electric energy was used were carefully observed. It was noted that the prevailing habits of city employees were continued, resulting in the operation of electric lights during the daytime in places where they might, with due regard to economy, have been discontinued, but while some directions were issued, which had the effect of some greater degree of care in this matter, the prevailing use of energy may be regarded as practically the normal situation in municipal buildings, which would probably have continued had the energy been purchased or produced in any other manner.

The determination of the cost of the production of electricity during the test involved very considerable care, and comprised a large proportion of the attention and expense involved in the trial. The plant itself has already been described, as well as the space in which it was installed. The methods under which it was operated were those which had been maintained up to that time, very careful attention being devoted to the most economical operation of each of the engines, using the larger engines when a sufficient load existed, and smaller engines at times when the load was decreased.

In order to effect as extensive a use of the wasted steam as possible, which would then become a credit to the cost of the operation of the engines and would be charged to the cost of the heating of the buildings, the operation of the smaller engines was continued during the night time, and as all the other machinery was discontinued at that time, the result was considerably to the advantage of the economy of the electrical generating plant. The storage battery was operated for periods which would apparently contribute to this general result, and was also found useful during short periods of repair required in the engine plant. The battery, however, proved to be of little economical value, as it involved a wastage of 12,267 kilowatt hours, and its useful effect was, therefore, only 50 7/10 per cent. of the energy which it absorbed.

The methods applied to the observation of these engines and generators were unusually complete and careful. A recording meter of the St. John type was connected in the steam supply of each individual engine, and marked on a paper tape the rate of flow at all times. A record was also maintained showing the period of time during which each engine was in use and the amount of steam absorbed in proportion to the electrical output of each engine was thus carefully ascertained. It was found that the steam used by the combined plant, for the entire year, was at the rate of 50 9/10 pounds weight of steam for each kilowatt hour generated, which represented the consumption of 6.45 pounds of coal, but to this, of course, had to be added that proportion of steam condensed in the pipes leading to the engines, and also the steam that was used by the auxiliary appliances required in the operation of raising steam for the engines. These items added nearly 20 per cent. to the fuel which was consumed, making the total 7.78 pounds of coal per hour, per kilowatt produced, or over 10 pounds per electrical horsepower.

The total expense of this service was determined by giving credit to these engines for the full expense of so much of the steam which they used, as was found to be utilized in the work of heating the buildings. This amount was ascertained by considering the contribution of the elevator engines and other lesser appliances, as having formed the first contribution to the heating work, the remainder being that credited to the electric engines, and forming the portion which would have to be supplied by high-pressure or "live" steam if the engines should be discontinued and the electric supply purchased from the public service.

The engines were found to have actually received during the year a total weight of 36,275,-352 pounds of steam, which was 44.43 per cent. of all the steam that was generated.

The proportionate share of the steam condensed in the pipes and the auxiliary pumps and appliances, was found to be 7,620,855 pounds weight of steam, so that the engines really absorbed 43,896,207 pounds weight of steam, or over 50 per cent. of all the steam raised. Of this amount the wasted steam from the engines which was used in the work of heating, was about 26 per cent. Even this result, restricted as it is, was only brought about by the abnormal addition to the requirements for steam in the heating season of the three exterior buildings, which had added to the actual work of the Hall of Records building, for which the plant was originally designed, about 175 per cent.

It is doubtful if even this moderate result would have been secured had it not been for the practice maintained in these buildings of continuing the supply of heating steam during the night hours. This afforded the opportunity of using the small electric engine during the entire night hours, and exhausting its wasted steam into the house-heating system, but in commercial practice it has been found that the heating of such buildings is more economically effected by discontinuing the supply of steam during the night hours, allowing the buildings to cool down, and then restoring the heat by a new supply of steam in the early hours of the morning.

Under such methods many buildings are now being heated by the use of steam during only 3,000 hours out of the total 5,000 hours of the heating season, but where it appears to those engaged in operating such plants, to be an advantage to the apparent economy of the production of electricity to create a means for the use of wasted or exhausted steam, as in this case, the process of night heating is carried on, although its value from the point of view of economy in the cost of heating is very dubious. In other words, by creating the wasteful condition of one service, an apparent economy may be effected in another service, and a fictitious appearance of reduced cost may be created by the process, which is similar to that of filling a vacancy in one pocket with coin extracted from another.

Since one of the particular objects of the trial was to determine the effect upon the cost of operation of the cessation of the use of these engines from the combined service of the plant, leaving the remaining apparatus in operation and contributing as far as they were found to have been adding to the service of heating, the results of the apportionment of the steam contributed by the engines to the work of heating, are not in any respect unfair or detrimental to the real expense of the maintenance of this electrical plant.

The other expenses which were involved by the operation of this plant consisted of labor, supplies and repair. The two latter were readily determined from the Departmental accounts. The labor employed about this part of the plant included the chief engineer, three watch-engineers and three oilers. The question of how much of this force would be necessarily retained if the operation of this part of the plant were discontinued, was one which presented some difficulty of decision.

It was generally agreed that a substantial part of the expense of this force was directly involved in the operation of the electric machines, but some difference of opinion arose as to the question of dispensing with the services of some of the four engineers. The Department representatives insisted that it would be necessary under any circumstances to maintain a licensed engineer on the plant on each eight hours of service, whether the electrical engines or any machines were in operation or not. It was, however, generally conceded that the three men employed as oilers, whose time was practically wholly employed by the engines, could be dispensed with if the engines were not in operation, which would effect a reduction of $3,295 a year. It also seemed probable that a less rate of wages could be paid to a chief engineer, in charge of a reduced extent of machinery, which would have had the effect of reducing the labor bill by a further sum of about $365, or a total of $3,660.

The final enquiries which were brought out by careful investigation in regard to the operation of this plant related to the fixed charges which the city was involved in by reason of the purchase of this machinery. It was found that the cost of the plant had exceeded $60,000, and that the present value of the plant was represented by a sum of $50,655. A rate of depreciation was also carefully determined, and found to average 5½ per cent. per annum.

These facts and figures thus afforded a view of the two sets of circumstances relating to the operation of this plant:

(1) The total cost in which the city had become involved by its investment in this machinery, as represented by the amount of energy produced that would have to be purchased from an outside system of supply.

(2) The cost at which the energy was now produced, assuming that the fixed charges upon the investment were a continuing expense in the event that the machinery should be shut down and its further use discontinued.

These results are shown in the following statements:

Operating expenses according to the Department view of the labor cost:

A City of Wasted Steam.

a Production cost of the steam used after deducting the exhaust used in house-heating	$11,154
b Supplies and repairs, as charged in the accounts	786
c Engine-room labor, charging only the three oilers	3,285
Supervision on items *b* and *c* @ 6.38%	261
	$15,486

Without any fixed charges, this results in cost of the production of energy of 2.17 cents per kilowatt hour, as compared with a purchasable price of 1.927 cent.

But this figure by no means represented the full cost to the city of electric energy produced in its plant.

The total expense to the city included:

Interest on the investment	$1,773
Depreciation	2,830
Rental value of space occupied by the engines....	2,636
Share of the fixed charges and space of the boiler plant	3,722
The operating expense according to the Department's view as above stated	15,486
Extra expense on Brooklyn Bridge service......	279
	$26,726

Making the cost per kilowatt hour as produced, 3¾ cents per kilowatt hour.

If the electric plant were charged as it should be with its fair share of the whole expense of engine-room labor, then the total cost would become $29,260, and the cost of the energy as produced, 4¼ cents per kilowatt hour, and of the net utilized quantity, 5 cents.

The division of the cost of electric production is as follows:

	Per Cent of Electricity	Share of Cost
Hall of Records	42.48	$12,430
City Court Building	4.43	1,296
County Court House	18.12	5,302
City Hall	10.10	2,955
General Power	4.34	1,270
Cellar Lights and Fans	17.10	5,003
Storage Battery Operation	3,43	1,004
Sewage Compressor (partial)		
		$29,260

The price at which electric energy could have been purchased from the public service system under the rates prevailing prior to the year of the trial, was 3 cents per kilowatt hour. This was later modified by a new and reduced scale of rates, which came into operation during the year 1914. This provided that over and above a certain quantity of electricity the energy supplied would be charged at the rate of only 2 cents per kilowatt hour; therefore, at the end of the trial it became apparent that any quantity of energy which would have to be purchased if the electrical generating plant were to be abandoned, could be secured at the price of 2 cents. To this should properly be added the charge for supervision at the rate of 6.38 per cent., making the cost 2.127 cents. But upon this rate the return of taxes to the city was found to be at the rate of 2/10 of a cent per kilowatt hour, thus reducing the cost to the city to 1.927 cent per kilowatt hour.

Stating the figures in totals, the cost of purchasing the required supply of 594,000 kilowatt hours, would be:

For energy at 2c	$11,875
Add supervision	745
Total	$12,620
Credit return by taxes	1,187
Cost of New York Edison Company's service	$11,433
This is to be compared with the Department's figures of operating expenses only	15,486
and the total expense, including fixed charges, of	26,726
And the real total cost, including a proportionate share of the engine-room labor, of	29,260

These statements very clearly demonstrate:

First. That the cost in which the city was involved in establishing and operating this plant was not a source of economy, but rather of continuing loss.

Second. That, after the mistake had been made and the investment had been sunk in the apparatus, it would nevertheless have been profitable to the city to cease the use of the apparatus, and to purchase the necessary supply of energy from the public service.

The extent of the annual loss to which the city was exposed during the period of the test was at least $15,000 according to Department figures, and nearer $18,000 a year, if all expenses are properly considered. This could have been saved had the plant never been installed. This result very plainly demonstrated the fallacy of the policy of installing such expensive plants in city-owned buildings, in view of the continuous decrease in the cost of purchasable energy, and the continuous increase in the expenses of operating such a plant, and particularly in the cost of labor attendant thereon.

The conclusion is enforced by the developments in additional expense since the period of the trial. The price of fuel has risen from the average of

$2.90 to a figure in excess of $5 per ton. The cost of the electricity generated has been increased by this item $4,400, while the price of purchasable electricity has remained stationary. The operating expense of the production of electricity would thereby be increased nearly thirty per cent.

CHAPTER X.

THE COST OF ELEVATOR SERVICE.

THE five passenger elevators in the Hall of Records were installed at a time when a change was developing in the methods of operation of such appliances, and when the tendency of improvement was all in the direction of the use of electricity, rather than water pressure, as a means of more effective service. The developments that have taken place since that time have demonstrated that the electrical elevator in its modern form has overcome difficulties which had been experienced in the early period of development, and has become not only more economical, but much more effective in carrying any given amount of traffic. This is due to the fact that the electric motor has the capability of developing power in proportion to the resistance with which it has to contend, and when applied to the movement of a car containing variable loads of passengers, it will operate at practically the same speed when the car is full as it will when the car is only partially filled. In particular, its capabilities in this direction are of great advantage in the saving of time which it effects in starting the car from a standstill and bringing it up to its proper speed.

The hydraulic elevator is provided with a sufficient power for this purpose, but it is not capable of any increase to meet an excess load, and moreover, the movement of the water through pipes, passages, valves and cylinders at high speed, sets up a degree of friction which retards the movement of the car and renders the whole combination relatively sluggish. This is most noticeable in the case of what is known as the plunger elevator, a form that became very popular at one time on account of its apparent security in the form of the plunger or steel rod under the car, extending down into a long cylinder let into a hole in the ground.

The ordinary observer not unnaturally assumes that this rod is of solid steel, and is capable of sustaining the car if anything should occur which might cause it to lose its support in the shaft. As a matter of fact, the rod is quite flexible, being composed of piping screwed together in short lengths. It is closed at the bottom end and actually floats in the water inside the cylinder, and the elevator is mainly supported by the ropes attached to the top of the frame, which are connected to an enormous counterweight, nearly heavy enough to raise the plunger and car when empty. The pressure of water when applied to the bottom end of the plunger is just sufficient to balance the rest of the weight of the empty car, and the load of the passengers inside the car. In this condition the weight pulls the car and passengers upwards, but the great weight of the ropes and the plunger, together with the water following the plunger, constitute a mass of material, which has to be set in motion and then speeded up to the full rate of motion of the elevator and then retarded by gravity to a state of rest, making a very cumbersome and sluggish combination.

As a result, elevators of this type will not give as much service in the handling of traffic as will an electrical elevator of similar proportions.

The work of any elevator mainly consists in stopping and starting, and the electric machine has superior capacity in this regard over any hydraulic apparatus.

The successive improvements which have been made in electrical machinery have also substantially reduced the amount of electric energy required to operate elevators, and such machines as those in the Hall of Records, which stop at every floor, could be depended upon at this time to operate, with a fair amount of traffic, upon a consumption of electricity of approximately 4½ kilowatt hours for every mile of travel by the car.

This situation had been so far recognized by the city authorities that when changes were made in the building of the County Court, the hydraulic elevators in that building were removed and two electric elevators were substituted. It happened that these were planned to be supplied by electric energy derived from the system of the New York Edison Company, which for such work is at a pressure or tension of 240 volts, and thus it was found that when the test was contemplated, that they were not suited to be supplied by the energy generated in the Hall of Records' plant, the pressure of which was only 125 volts.

This is a situation which frequently occurs in connection with such mixed installations of power and lighting apparatus, and in several city-owned buildings has brought about very confusing conditions, whereby part of the plant and the lighting may be provided by generating machines and engines, while the remainder is necessarily operated by purchased energy.

During the trial, therefore, the two elevators in this County Court building were operated by electricity purchased from the New York Edison service, and it was found that they had used during the year of the trial, about 45,000 kilowatt hours, which would have added about 6 per cent. to the total output of the plant if it had been served thereby.

In designing the equipment of the Hall of Records, it would have been much better from the point of view of an economical operation, if electrical elevators had at that time been installed, which, with some modernized improvements, might have well been served by the generating plant, and would have eliminated from the plant the rather inefficient apparatus necessary for pumping water for the hydraulic machines. On the other hand, the two elevators in the City Court building were also of hydraulic pattern, and it was found that these could be operated from the Hall of Records, and pipes were connected for that purpose, so that the plant in the building during the period of trial, operated the five passenger cars in the Hall of Records, the two cars in the City Court, and the ash-hoist in the boiler-room. Four hydraulic book-lifts had been provided, one in each corner of the building, the intention of which was to operate dumb-waiters to carry documents from floor to floor. These were found to be so entirely unpractical that they have never been used during the entire period of their installation.

The pumps which supply such hydraulic installations are usually of the form which is known as a direct-acting, reciprocating pump, and they are usually of what is known as the "duplex pattern." This form of pump is very simple and inexpensive in first cost, but under customary conditions as the pump becomes worn, the tendency increases towards wasteful conditions by reason of the shortening of the movement of the pistons and water-plungers, under which condition the piston in the cylinder does not quite complete its stroke, and steam is required to fill this space at every reversal of its movement.

Furthermore, when this kind of pump is utilized in connection with a house-heating system, any rise in the pressure in the heating system will operate in this same direction and will further tend to decrease the movement of the piston and increase the wasteful operation. It thus becomes very important with this kind of pump, to reduce the pressure required for heating buildings with exhausted steam, and as in the case of the electric engines, apparatus is necessary to facilitate the circulation of the steam in the radiators and the piping, and prevent the pressure rising above a maximum of about a pound per square inch.

When these pumps are used in irregular elevator service, they suffer from another difficulty arising from very frequent stopping and slow movement, during which circumstances steam continues to be condensed in their pipes, passages and cylinders, and the total consumption related to the work which is done, becomes unfavorably expensive. Thus under favorable circumstances, it has been found that the combination of such pumps with hydraulic elevators of standard form would require approximatelv 550 pounds weight of steam for every mile traveled by the cars of the elevators, but in the case of the Hall of Records, as probably in many others, it was not surprising that the trial demonstrated a considerably larger use of steam, which during the year was found to average 796 pounds per mile of car travel, an addition of approximately 50 per cent.

There were times when for a period of a week this figure was further increased to a maximum of 960 pounds of steam per car mile. Such results are not unknown to those familiar with the operation of elevators, but it had not been practicable prior to this trial to determine the further expenses attached to such pumping plants by the cost of operating the auxiliary appliances which provided the means of the steam supply, and also the loss of steam in condensation in the piping leading to the engines.

During the trial this interesting point was very

definitely determined, and it was found that these additional demands added 25 per cent. to the average amount of steam consumed, and brought the total for the year to an average of 1,004 pounds weight of steam for each mile of car movement. Such a fact very simply and readily determined the inefficiency of the apparatus from a monetary point of view, because it was evident that the mere cost of the fuel required to raise the steam amounted to 19 cents.

The test observations determined the quantity of this steam which had been reused in the heating of the buildings, and which could, therefore, be deducted from the cost of the operation of the elevators, and charged against the cost of house heating. This was 6.19 cents, which left the net cost for fuel only of a car mile of travel, at 12.88 cents. Inasmuch as an electrical elevator would have operated the same service at 4½ kilowatt hours per car mile, and this could have been purchased at the time when the trial was started, at a cost of 3 cents (and later for 2 cents), it will be seen that the price of coal alone was practically the same as the cost of purchased electricity, without regard to the labor and upkeep involved in the operation of the plant, and of the generating of the steam by which it was operated.

The elevators in the two buildings were found to have traveled during the year of trial a distance equivalent to the movement of one car over 16,424 miles. Such a figure may seem surprising to those who do not realize the immense distances covered by the movement of elevators, which in the City of New York vastly exceed the entire travel of railways and surface cars within the area of the Greater City.

The trial brought out the actual operating costs of this part of the plant in very definite form, the details of which are shown in the table on page 81. The actual cost of operation, inclusive of steam, supplies and repairs, and that portion of the engine-room labor charged in the departmental accounts to this service, aggregated $4,999.22, to which must be added the share of other labor and supervision, bringing the operating expense, without any fixed charges, to $6,859.32, or over 41 cents for each mile of car travel.

The fixed charges upon the installation of the pumping machinery, tanks and piping, amounted to a further sum in excess of $3,379, so that the total cost of the operation of the elevator plant thus involved an expenditure of $10,238, or 62 cents for every mile of car travel.

The operation of these elevators by the electrical method could have been effected by electricity generated by the plant, by utilizing the storage battery to neutralize the effects of the sharp demands for energy. At the total expense of 4¼ cents per kilowatt hour, the energy required for the elevator operation would have cost $3,140.

There would have been no need for steam, nor for any pump and other appliances, so there would have been no current or fixed expenses, nor labor on those items. Supplies and repairs, with labor thereon being estimated at $1,500, the total expense would have been $4,640, or about 28 cents per car mile, or less than one-half the cost of hydraulic operation. If operated by energy purchased from the public system a still more economical result could have been secured, the expense for current being reduced to $1,400, and the total operating expense to $2,900, or 18 cents per car mile.

The installation of the hydraulic form of elevator was thus clearly a mistake of judgment at the time of the construction of the building, but it did not constitute the same kind of error as the installation of the electrical generating plant, because the elevators were a necessity to the building, and the only question that required careful prevision was the method of their operation. If careful foresight had been devoted to this part of the subject, it would have been considered that with the declining prices of electrical energy provided by a public service, or even with the installation in the building of an electrical generating plant, the probable future economies of the elevator service lay in the direction of electrical operation.

The mistake, however, having been made, the question that arises, in view of the investment having already been made in the hydraulic form of machine, is whether a sufficient advantage can be gained by the investment of a further sum of money in the reconstruction of these elevators in electrical form. In such case, whatever economy may be effected in the cost of operation, must be discounted by the fixed charges that would result from the expenditure of further capital upon the new apparatus.

In order to decide this interesting point, a special study and test was made to establish the maximum demand for power which the elevator service created, and this test decided that the largest demand for the seven elevators operated was not in excess of 192 horsepower in the form of water under pressure.

An estimate was then made by the manufacturers of the elevators of the cost of remodeling the whole of the elevators into the electrical form, and it was found that this would involve an expense of about $48,000. Making due allowance for the fixed charges of interest and depreciation upon such a new investment, it is found that the elevators in this form could be operated at a total cost of $7,800, a saving of about $2,450 a year.

Such a change would, moreover, effect other indirect economies. Besides a direct saving over the cost of the existing system, the removal of the pumps, and the reduction of the use for steam would eliminate the last requirements for maintaining the most expensive part of the labor force employed by the plant, if at the same time the electric generating engines should also be dispensed with. From every point of view, therefore, the result of these observations indicated that the operation of the hydraulic elevators in these municipal buildings was relatively expensive, and although any change for the better is hampered by the necessity for the expenditure of a large sum of money upon a new form of apparatus, it would be productive of economy.

CHAPTER XI.

WATER SERVICES.

BUSINESS buildings of the character of the Hall of Records require a number of minor services involving the use of machinery in connection with water. As has been explained, the deficiency of pressure in the public water supply necessitates the pumping of water up to the tanks located in the roof of any buildings, the height of which exceeds a few stories above the street.

In addition to this, part of the water so pumped has to be heated and circulated for use in the lavatories, and for cleaning purposes. It is necessary also, in the case of buildings provided with deep cellars, to pump out waste-water, and to lift sewage from any fixtures that may be situated below the level of the public sewers.

In the Hall of Records a part of the water which was pumped was also filtered, cooled and circulated in separate piping for drinking purposes involving the use of refrigerating machinery.

The pumping of water from city mains, which so long has been necessary in the City of New York, appears a strange accompaniment of conditions of sanitation in a great modern city, which has spent more than three hundred millions of dollars upon its water systems, only to provide it in so ineffective a form that it fails to reach the upper floors of buildings of moderate heights. The deficiencies in this respect are due to two causes: first, the inadequate height of the reservoir system supplying the down-town portion of Manhattan, and secondly, the inadequate size, and the complexity of the system of underground water piping.

These deficiencies were, of course, aggravated in their effect by the continually increasing demands upon the supply, largely due to the use of water for commercial purposes, and particularly for use in such power plants as that in the Hall of Records. The situation has only been modified by the new supply derived from the Catskill system in regard to one feature, namely: the height at which the reservoir is situated.

The same system of underground piping remains in use, much of it being of inadequate size so that the increasing demands for water reduce the pressure to a large extent and nullify the advantage provided by the additional height of the new storage reservoirs.

One of the fundamental elements in the provision of municipal water service seems to have been largely ignored in the operation of the water service of New York, and the facts disclosed in the trial threw considerable light upon the subject. A municipal water supply is provided at the expense of the community, as a necessary public service, or a general obligation for sanitary and fire protection purposes. Its use in commercial processes is for profit-making purposes, and the diversion of the public supply to such operations as the generating of power in private buildings, or even in municipal buildings, clearly goes beyond the general intention of the investment of public funds in a public necessity.

The regulations of the Water Department ever since they were formulated, prohibit the use of the water in the mains as a means of motive power. Water must not be taken from the city mains and used to set in operation an elevator or a water-wheel, or any other machine which may be directly operated by water pressure, but the practice has been allowed to grow up of using the same water in vastly greater amount to be evaporated in a steam boiler, utilized to drive machinery, and then discharged to the atmosphere, or condensed and run into the sewers.

The small utilization of power for any domestic purpose is therefore barred, but any person may take the same quantity of water or much more, turn it into steam, and use it for the production of a larger amount of power, without any restriction, and for no greater rate of payment than that which is paid for water which is absolutely necessary for sanitary purposes.

The most aggravated situation in this regard is the increasing use of water from the public mains for use in refrigerating apparatus. In this service the water is merely used for the operation of condensing or cooling the gases used in the process, and vast quantities are run over piping and through tanks, and are wasted directly into the sewer, at rates of as much as 12 to 15 tons a day for each ton of rated refrigerating capacity.

In many hotels, restaurants, stores and business processes such apparatus is operated for the purpose of decreasing expenses in the purchase of ice, and it is not infrequent to find machines of from 25 tons to 100 tons capacity, using from a hundred thousand tons to half a million tons of city water every year. The price which is paid, of one dollar per 1000 cubic feet for this profitable use of water in what is really a manufacturing process, is no more than that which is paid by the domestic consumer for the requirements which are forced upon him by the necessities of public sanitation.

Such excessive demands upon a water-system which was planned for and expected to be limited to necessary sanitary purposes, has resulted in reducing the available pressure at various points in the system to such an extent as to necessitate additional work in other buildings to pump the water required for the sanitary service.

The same excessive demand for water for such

purposes has brought about a far wider effect. The occasional shortage of water reserve, aggravated by the intensive demands for water for these commercial purposes, brought about an agitation which involved the city in the construction of additional water supply and storage capacity. It was mainly these circumstances which led to the enormous undertaking of the Catskill water system, which was represented as of urgent necessity at the time of its inception, and is now recognized to have been undertaken many years in advance of its real necessity.

The recent expenditure by the city of one hundred and thirty-five millions of dollars in the establishment of that water service, for which there is no new demand that was not capable of being served by the existing systems, may thus be attributed at any rate to a considerable extent, to the utilization of water in business, and for power purposes.

After the construction of this vast system was under way, economies were effected, by the elimination of leakages and wastages, which brought about a substantial decrease in the amount of water then required, which reduction was especially marked in the Borough of Brooklyn, where the use of water had been the most wasteful.

The new investment has, however, loaded the operations of the Water Department with a financial burden of fixed charges and expenses which has entirely absorbed all the profits that for many years were derived by the City from the sale of water, and has brought about a very heavy deficit, amounting at the present time to upwards of seven millions of dollars per annum. The effect is far-reaching, since the issues of the bonds for the purpose of water service, were heretofore self-sustaining, and were thus released from the constitutional restrictions upon the borrowing capacity of the City. It is now necessary to regard them as being incapable of sustaining their own fixed charges, and it seems doubtful whether they will regain that condition within many years, unless a higher price be charged for water, or some additional revenue be secured from the use of water for commercial purposes.

These facts bore directly upon the question

which arose in connection with the operation of this plant, as to whether or not the water that was used in its operation constituted an expense to the city or not. Investigation disclosed the fact that the several departments of the city made extensive use of water in various municipal operations, but never accounted to the Department of Water Supply for the expense, nor made any charge against their own operations for the water they used.

The result of the discussion, while it was of interest in bringing out these points, led to the conclusion that under this existing system the water involved no present expenditure on the part of the Department in charge of the building, and it was not therefore included as one of the costs of operation. It was, however, readily recognized that in the case of the purchase of steam from a public system, as well as in the purchase of electricity from the public service, the city gained a return by the water that was used in the processes of generating steam and electricity, and in considering the purchase of steam in particular, this return constituted a substantial reduction in the cost of the public steam service to the city.

The water that was used in the Hall of Records for general sanitary purposes in a year was about three million cubic feet, and would have cost, if purchased at the regular rates, $3,000 a year, and the water used and wasted by the plant of machinery was 662,000 cubic feet, worth $662. These costs would not have represented its entire expense, since the operation of pumping it, circulating it, and cooling part of it, had to be carried on.

The machines provided in the Hall of Records for pumping the water were of two forms: one being a steam-operated, direct-acting pump, and the other an electrically-operated motor-driven pump. Small steam pumps of this character are notoriously wasteful in their use of steam, and it therefore became a matter of considerable interest to secure during the trial, a determination of the actual amount of steam used in relation to the water pumped in these buildings, and to compare these with the cost of operating an electrically-driven pump.

It was found that the water, in order to be

Waste Steam in the Public Sewers.

pumped to a height of about 110 feet, involved a pressure at the pump, equivalent to 200 feet, the increase or difference representing the friction created in passing the water through the pipes and bends in the piping. Steam was consumed in this work at the rate of 266 pounds for each horse power represented by the water lifted. This involved a consumption of 33 pounds of coal per hour, the value of which alone was four and three-quarter cents for each water-horsepower exercised over a period of one hour.

The full expense of the service during the year included the following items: The cost of steam used (after crediting that utilized in house-heating), with supervision, was $1,061.17; the share of the engine-room labor, supplies and repairs, with supervision, was $456.16, and the fixed charges added $463.00, making $1,980.33 in all.

This represented an addition of 54 per cent. to the value of the water which was used, and a cost of nearly 10-4/10 cents for each water horsepower hour. An electrical pump could have been operated by purchased electricity for an expense of $730, and adding the same share of labor, supplies, and repairs, or $456, and with interest and depreciation continued as at present, $108, the total cost of electric operation becomes only $1,284, which is less than seven cents per water horse-power hour, and seven hundred dollars a year less total expense.

These figures make it clear that the process of pumping house-water by steam power is not as economical as it would be with the use of a pump operated by electricity, even though the steam pump provides some part of its steam for heating purposes.

The expense of operating an electrical pump may be substantially reduced if the apparatus be provided as is customarily done, with an automatic controlling device, thus reducing the attention required to an occasional cleaning and oiling of the machine. Under such circumstances the expense for labor, supplies and repairs which was incurred in the Hall of Records process, could have been substantially reduced, and it is probable that the same amount of water could have been pumped in that building for a cost, including the fixed charges on an electrical installation, of less than $1,000 a year, or practically one-half of the ascertained cost of the steam operation.

As there was an electrical pump installed in this building which had been placed there as a reserve, the opportunity was afforded of making a direct comparison between the expenses of the two systems, but unfortunately the gearing of this pump, which was of a somewhat antiquated pattern, was broken by accident, and repairs were not effected by the Department during the rest of the period of the trial.

This little incident throws considerable light upon one of the fundamental deficiencies of municipal operations in connection with machinery, namely, the delay in providing for necessary repairs, and even for necessary upkeep in advance of accident or failure of the machine. This is due to the very methods of municipal reform which have been set in operation to protect the city's finances from unregulated expenditures. It is, under this system, necessary to secure an appropriation, or to make provision in a budget for repair work, processes which occupy considerable time, and in the case of a budget cannot be always foreseen.

Thus necessary repairs are deferred until an accumulated amount has been reached which renders it worth while to secure an appropriation, and in the meantime the operation of machinery is conducted at a disadvantage, and often at a loss to the taxpayers.

A portion of the water, about one-tenth of the whole quantity drawn from the city mains and delivered to the house tanks, was warmed for use in the sanitary systems of the buildings. Part of the warm water in business buildings is required for the purpose of scrubbing the floors, but a greater quantity is used in the lavatories. The quantity of water used for that purpose is found to be very variable, and fluctuates according to the methods under which it is circulated, and also doubtless by the habits of wastefulness of the occupants.

It is difficult to say which of the two is productive of greater wastage. Waste exists under any conditions, and it is not only a loss of the value of the water, but of the cost of its pumping, and the cost of its heating and of its circulation.

Where efficient circulating arrangements are lacking, it is the custom of those who use the supply of water to draw off water until the supply runs hot, and thus a considerable quantity of water, which has previously been heated and has lost its temperature by lying idle in the pipes, is thrown to waste at every use of a hand-basin by an occupant.

If an effective circulation is provided, as was done by mechanical means in the Hall of Records, then this wastage of the water may be greatly reduced, but on the other hand, a much larger amount of the heat, which has been absorbed by the water, is lost in the process of circulating the warm water round and round in the system of piping. Moreover, the cost of that circulation becomes an addition to the expense of the process.

In the Hall of Records the pump which circulated the water was operated by electricity. The steam that was used to heat the water was in part exhausted from the power apparatus, but of course, the full value of that steam is charged to the cost of heating the water.

It was a notable fact in regard to the use of warm water, that even in the summer months there was little reduction in the use of heated water. This was doubtless due to the fact that the larger part of the heat that was absorbed by the water, was lost in the process of circulation.

In heating the water 1,548,050 lbs. of exhausted steam and 334,069 lbs. of high-pressure or live steam were used. The cost of producing this steam was $647.

The water which was heated was approximately 300,000 cubic feet. The bare cost of warming this water was thus more than two dollars per one thousand cubic feet, or twice its original value. Adding the expense of the operation of the electrical circulating pump, it appears that a supply of warmed water costs about $3.50 per thousand cubic feet, or three and a half times as much as cold water.

Some economy may be effected in the Hall of Records by the more simple system, such as that by which the water was originally heated in these buildings, of using separate heaters, supplied by purchased steam, the water being circulated by gravity, thus reducing the losses involved in the long system of circulation, and in the expense of the operation of the electric pump.

Although the service afforded by warm-water appears expensive, it assumes a relatively economic aspect compared with the expenses involved in discharging sewage and waste-water out of such a building as the Hall of Records, which service answers no economical purpose whatever. It is in point of fact, an obligation wholly resulting from the fact that the sub-basement in this building, constructed for the purpose of housing the machinery, was necessarily far below the level of the public sewers.

Such spaces are in themselves, relatively expensive, especially under such circumstances as their construction in the rocky soil of the Borough of Manhattan, where every part of the space has had to be blasted out, and where every part of the walls and columns is necessarily of the largest and most expensive proportions.

In the absence of a machinery plant, the use of such a space might have been dispensed with, but when constructed for this purpose, it must be rendered habitable, and its use by employees involves the installation of sanitary apparatus, also provision for dealing with the breakage of piping, or possible flooding of the exterior vault spaces which provide the sub-basement to some extent with air and light.

In this building, as in a number of others, the drains leading from these sources of rejected water are led to a pit sunk below the floor, in which the sewage, etc., is collected alternately in one of two closed receptacles, each having a capacity of sixty gallons. These are automatically discharged by admitting compressed air to the interior of the receptacle, which presses out the collected sewage through a pipe to the public sewer. The air under pressure is provided by steam-driven compressors, which are customarily very wasteful in operation. Measurement applied to this compressor during the first thirty-two weeks of the year, demonstrated its average use of 4,620 pounds weight of steam per day, which was equal to the continuous operation, day and night, of a boiler of six horsepower capacity.

The Department then decided that an electrically operated air-compressor would be more economical, and purchased and installed such an appliance, which was tested and utilized after the thirty-second week to the forty-eighth week. An interesting comparison was thus afforded between the cost of steam power operation and electrical operation for equal service, and it was found that the average amount of electric energy which was required for each 6,000 gallons discharged by the sewage-ejector was 4.32 kilowatt hours. The cost of operation of the steam system was found to be as follows:

Covering the period of test during which 4,667,100 gallons, or 77,800 cubic feet of sewage were discharged, the steam (after allowing for such as was utilized and charged to house-heating) was 733,310 pounds, the cost of production being $254.95, or $3.30 per 1000 cubic feet which is to be compared with the cost of 3360 kilowatt hours of electricity at 2 cents, which was less than $1.00 per 1000 cubic feet. The cost of steam alone was thus at least three times as expensive as that of purchased electricity would have been.

Labor, supplies and repairs, with supervision added $324.42. The fixed charges of the apparatus and space it occupies, are $635.00. The total of $1,214.37, thus represents an approximate cost of $15.60 per 1000 cubic feet.

A change to electric operation under existing circumstances would be of economic advantage, by the purchase of an electric pump, which is an effective appliance for such purposes, and is in use in many other buildings.

The entire expense of operation of such an apparatus, even with the existing charges for labor, and for interest and depreciation on the abandoned ejectors, would be only about $730, or 40 per cent less expense than that of steam operation.

The cost of this unremunerative service of sewage removal represented a daily expense of $3.33 and for each man employed in the machinery and fireroom space an annual expense of eighty dollars.

CHAPTER XII.

THE COST OF REFRIGERATED DRINKING-WATER.

ONE of the refinements of convenience which has been provided in some commercial and residential buildings, is a system of water supply for drinking purposes, which during that portion of the year when the public supply has risen in temperature, is cooled by artificial means. This service could, of course, be conducted by the simple process of purchasing ice and circulating the water through pipes in contact with the ice, or more simply still, by the usual process of ice placed in water coolers. But complicated machinery is often adopted without regard to alternative and simpler methods of securing equal results, and thus some buildings are provided with an installation of expensive refrigerating apparatus, by which the cooling effect that could be equally well produced by melting ice is produced by steam power.

The process in the Hall of Records involved the use of steam to evaporate ammonia into a gas. Cold water is then used to cool the gas before it is allowed to expand in a system of piping, the expansion resulting in an absorption of heat from the pipe surfaces, whereby a refrigerating or cooling effect is produced. A steam pump is used for returning the ammonia to the boiler or evaporator. The process is thus quite elaborate and involves considerable investment in the installation, and care and expense in its operation.

The extravagant use of water which is made in the process of cooling the gases, has already been referred to. The value of the entire operation must be limited by the purchasable equivalent in the form of ice.

Refrigerating apparatus is largely used in hotels. In some it has been found that their expense is much in excess of the cost of an equivalent amount of ice, but the machinery is maintained on account of the convenient form in which the cooling effect is distributed, and the various degrees of temperature that can be secured. Thus in certain processes, such as making of ice-cream, very low temperatures must be maintained, but in some other culinary processes ice would answer just as well. The expense of machinery, therefore, is really chargeable to some special process such as the manufacture of ice-cream, which might be purchased at less cost. So far as drinking water is concerned, only a moderate amount of cooling effect is required, as the temperature of the general public supply only rises from an average of about 40 degrees to an average slightly over 70 degrees, and in the summer time requires to be cooled only from the latter temperature down to approximately 45 degrees Fahrenheit.

The habits of occupants of business and residential buildings in New York are adverse to economy in the use which is made of this convenience, and it has been found that excessive wastefulness prevails. Thus in one large hotel where the iced-water faucets were placed over sinks, it was found that the help were in the habit of using the cooled water in summer time for washing their hands, and even for cleaning tableware.

In most cases the water must be circulated constantly around an extensive system of pipes, in order that it shall not increase in temperature while remaining stagnated in the piping. Such losses are similar to those experienced in circulating heated water, but they are intensified in the case of refrigerated water, since the process of reducing temperature is greatly more complicated and expensive than the process of increasing the temperature of water.

Some of these facts might have been known to those who were responsible for installing the system of drinking water piping in the Hall of Records, but nevertheless the installation was made, and was arranged with features which contributed to the greatest extent of wastefulness, inasmuch as the iced-water faucets were placed over the hand lavatory fixtures. Moreover, the maintenance of the supply is not confined to the warm season of the year, but a circulating apparatus was kept in motion throughout the larger part of the winter, to circulate the water, which would otherwise become somewhat warmed by remaining in the piping.

The use of the refrigerating apparatus is limited, of course, to the warmer portion of the year, and in the case of the Hall of Records the appliances were in service only during a period of twenty-four weeks.

As soon as this apparatus was set in motion observations were made of the use by the occupants of the building of the cooled water, which threw very interesting light upon the character and extent of the wastage, and the limited value which such a service afforded, compared with the expense of its production. It was found that the employees in the building had discovered that by tying down the spring lever of the faucet on the drinking water supply with a rubber band, a flow of the cooled water could be continuously maintained into the wash-basin, in which they were in the habit of placing bottles of milk for their luncheons. It was ascertained that probably 75 per cent of all the water that was cooled was diverted in this manner from its intended purpose.

A comparison was fortunately available with conditions existing in a commercial building in the vicinity, in which a similar service is in use by a large number of tenants and a staff of manufacturing employees. In this building it was found that the use of the water was conducted in a much more economical manner. The water which was actually drawn from the system in this building represented 20 per cent of the total cooling effect, 80 per cent having been lost in the process of circulation around the piping.

In the Hall of Records only 5 per cent of the work of refrigeration was represented by the water withdrawn from the system, and 95 per cent of all the work effected by the apparatus was lost in the process of circulating the water around in the piping. Even this statement fails to indicate the extent of the extravagance of the whole process, for the amount which was drawn from the system in the Hall of Records was the equivalent of a consumption of 15,000 drinking glasses of water during a business day of about eight hours. This represented approximately thirty glasses of water a day for each occupant of the building.

The expensive character of the process of refrigerating by this class of machinery was very well determined by the records of the trial, and by a special test which was conducted during a period of five weeks. During these observations it was found that the capacity of the apparatus, which was 10 tons of refrigerating effect, or 5 tons ice-making capacity per twenty-four hours, was utilized only to the extent of 56 per cent, representing the equivalent of the melting of 2-8/10 tons, or 5,600 pounds weight of ice per twenty-four hours.

This was the cooling effect required to maintain the low temperature upon the whole body of water circulated around in the piping, and also to reduce the temperature of such additional water as was drawn from the house tanks to replace that which was drawn out of the faucets. The same effect of cooling the water could, of course, have been effected in receptacles for ice in each of the offices, in which case the actual work that was effected in cooling the water that was used, extravagant though that amount was, would have been accomplished by purchasing 700 pounds of ice a day, the cost of which would have been about $1.40. The only inconvenience accompanying that method, which is in common use elsewhere, is that of the introduction of ice into the building and its carriage up to the point of usage.

Under the circumstances described, it will not be a matter of surprise to learn that the work of maintaining this drinking-water service was very costly.

The cost of steam, labor and supplies aggregated $1836, and with all expenses and fixed charges the service involved a total of $2,798, to which should be added the cost of water used in condensing, about $200, making a total of nearly $3,000.

Ice could have been bought, equivalent to the total cooling work which was effected, for a sum not exceeding $1,600, and the City would have been saved the investment in this plant, of about $7,500.

In the actual use that was made of the service in the form of drinking-water, the supply of ice for the season's service could have been purchased for about $450.

Allowing two glasses, or a pint of water, as a reasonable amount to be consumed by each occupant daily for every working day in the year, the expense of the system for an average of 600 occupants in the building was about $1000 per one thousand cubic feet of water, or about eight-tenths of a cent a glass.

While this refrigerating system was in operation, its evident extravagance led the departmental officials to adopt a method whereby the cost of production might be somewhat reduced through the use of exhausted or waste steam, instead of the live steam for which it had been planned and installed.

In order to effect this purpose a change was made in the piping of the pumps and appliances in the room in which the refrigerating apparatus was situated, the effect of which was to close up the outlet of the wasted steam from some of the pumps so that the steam would be forced into the refrigerating apparatus under a pressure of about 15 pounds to the square inch. This pressure is necessary, in order to properly effect the process of evaporating the ammonia, but it of course, had some reflex effect by retarding the operation of the pumps, and by increasing their demands for steam, so that it was not entirely an economical gain.

Moreover, the larger part of the steam used by the apparatus was still required to be high-pressure steam which was necessary to operate the ammonia pump or compressor. The little experiment, however, led to much wider developments in an unexpected direction.

The imagination of some of the departmental officials was fired by this apparent economy, and expanded into the conception of a large plant for the manufacture of ice, to be operated by the exhausted steam of all of the electric engines and the pumping machinery. They went so far as to suggest that an ice-manufactory large enough to utilize the exhaust steam during the summer should be installed in the Hall of Records, regardless of the character of the building, and of the danger attending the introduction of ammonia-using apparatus in the lower part of a public building containing documents and records of priceless value. Nevertheless, this proposition was communicated to the Board of Estimate and Apportionment, and a resolution was introduced in that body proposing that steps should first be taken to utilize the existing apparatus in the Hall of Records to manufacture ice, as well as to conduct its existing duty of cooling water.

This statement was supported by the assertion that twenty-five tons of ice a day could thus be manufactured, which of course, was absolutely impossible, as the full capacity of the machine was only five tons a day, and of this, 56 per cent was already in use upon the work of cooling water. Moreover, the apparatus had no means and no appliances for the manufacture of ice, for which it was neither planned, nor designed, nor was there any space available in a suitable position for such a purpose.

Notwithstanding all these facts, statements were given out to the press that ice could be made under such circumstances, at a cost of $1.50 a ton, and an application was actually made to the Board for an appropriation of $40,000 for the purchase of ice-making apparatus, which dubious experiment might have been carried out had not the facts brought out during the test, as to the real circumstances surrounding such a proposition, been brought to the attention of Mayor Gaynor.

At the prices of fuel at the time of this discussion, the items of coal, labor and supplies alone amounted to from $1.25 to $2 per ton of ice in well-arranged and economical factories manufacturing ice in large quantities. It has always been the case that ice is manufactured on a very narrow margin of profit, which is invariably restricted by the competition of the natural supply. It was naturally absurd to suppose that ice could be produced under municipal conditions, and at the rates of city wages, at prices competing with commercial rates. The cost of steam, of labor and supplies in the Hall of Records plant was about $3.50 per ton of ice-making capacity.

These points appealed to the Mayor, who responded to communications sent to him in the following characteristic letter:

CITY OF NEW YORK
Office of the Mayor

June 10th, 1913

Dear Mr. Bolton:

I am not surprised to receive your letter about the proposed new city ice plant. The statement given out is that certain engineers approve the enterprise. The Board of Estimate and Apportionment never appointed any engineers to examine into such a matter, that I am aware of, and I never heard of any reports to that Board. You say that you were appointed by Borough President McAneny, with other engineers, to look into the matter, and that you are convinced that the city could not profitably manufacture ice. If the city is going into the manufacture of ice for the poor we ought to do it after a committee has examined thoroughly into the matter. But you know an election is coming on and the matter is deemed one to catch votes with, and some people do not want to wait for a careful investigation. And I must say that I do not understand that there are so many poor people asking us to make ice for them. As soon as the matter is mentioned you hear it said by officials and others that the poor people on the East side must have ice made for them. I think I am as familiar with the East side as any member of the present city government. I have taken occasion to look all over the East side. I deny that the people on the East side are paupers. On the contrary, I find them to be an industrious well to do community. The percentage of pauperism among them is small. This continual libel on the East side of their being paupers and criminals is annoying to those who live there, and to a great many other people. I happen to know that there are very few people that are asking that the city provide them with ice. During the last three years I have advertised to the public through the newspapers that anyone who is too poor to pay for ice at the prevailing rates may go to any of the charitable societies, or head-quarters of the settlement workers, and obtain ice tickets without charge. I expect to do the same thing again in a few days. I am not at liberty to reveal the name of the person who provides these tickets. I hope he will let me reveal his name this year. But the quantity of ice that is taken in this way is extremely small.

It would surprise you to learn how small it is. I have not the figures in my mind just now, but if you are interested to know them I will have them sent to you. Let us have an end of all this libeling of the people of the East side by talking of their pauperism, and of their need for the city to furnish them with free ice and free this and free that. There are some people very solicitous about the condition of the people on the East side, especially one day each year, namely, election day.

Very truly yours,

(signed) W. J. Gaynor,

Mayor.

In a later communication the Mayor stated that about 900,000 pounds of ice were given away free during the year 1912, through the agency of the Department of Health, the whole of which was a free gift on the part of the Knickerbocker Ice Company, which has conducted this charitable undertaking for a number of years past.

It was a rather interesting comment upon this discussion that an investigation was made at the time into the circumstances of a municipal ice-making plant which is operated on North Brother Island, chiefly by prison labor. The product of the plant in this not wholly savory situation, was found to be disposed of by shipment on the boats of the Department of Correction, to other city institutions. It was ascertained that during the process of transit, 40 per cent of the volume of the ice was melted, and approximately 50 per cent of the remainder would have been unnecessary under economical management.

These facts and disclosures rather pungently illustrated the fallacy underlying the hopeful anticipations of municipal management and conduct of such commercial processes as the manufacture of ice, and throw a rather informing sidelight upon the point to which such enthusiasts might better direct their attention, namely: the economizing of prevailing wastages in the use of expensive products and conveniences.

CHAPTER XIII.

THE COST OF PUBLIC SERVICE.

IT was one of the anticipated effects of the test of the power plant in the Hall of Records that it would provide the basic facts wherewith a direct comparison could be made of the cost of operation during another year of trial. During this second period the services of the building, to the same extent and of the same character would be operated by means of the public supply of electricity, either with or without the use of the public supply of steam. The figures of cost determined during the first year of trial were to afford the means of deciding the question of the comparative value of the continued use of the boiler plant, or of the purchase of steam. This anticipation was embodied in an agreement under the conditions of which the trial was undertaken by the three parties in interest.

The expectation that this comparative trial would be undertaken added greatly to the interest of the observations conducted during the year of test, and had it been carried out it would have afforded a very clear demonstration of the methods, and would have settled any and all questions as to the details of the change of system. It was not, however, necessary to conduct such a trial to decide the relative cost of the supply of electricity and steam, because the facts obtained during the test provided complete information as to the quantities of steam and electricity which were generated, and those which were utilized in each service.

There was practically no element which afforded any opportunity for divergence of opinion as to the effect of discontinuing the use of the generating plant, except that of the extent of labor which would be found necessary under the changed and reduced operations.

The Department conceded only a reduction of $2,944 in wages, and the engineering advisers of the Bureau of Municipal Research estimated the saving at $4,683. Both of these fell short of what could readily be effected if economy in this direction were sincerely desired.

The other question which was involved concerned the quantity of coal that would be required under these changed conditions, a question which also would have been definitely settled by the comparative trial. But there was no wide divergence of value in this matter. The engineers representing the three several interests, made separate estimates, some of which were the results of an extraordinary amount of detailed study. Their estimates varied from a maximum of 4054 tons of fuel (or 68,600,000 pounds of steam) as estimated by the Department engineers, to 3585 tons of fuel (56,599,000 pounds of steam) as estimated by the engineers representing the Bureau of Municipal Research, and a minimum of 3,330 tons of fuel (52,300,000 pounds of steam), which was the estimate of the engineers representing the New York Edison Company. The largest difference was thus only 724 tons, the total value of which would have been about $2,143. The two other estimates varied only to the extent of a value of $754 which was quite an immaterial amount, in view of the large difference between the total expense of the operation of the plant

Burning Cheap Fuel.

and the purchasable prices of steam and electricity. The differences of estimate, such as they were, were largely due to the assumption on the part of the City engineers, that the wastages and losses which accompanied the operation of the plant, would necessarily be maintained after the operation of the generating engines had come to an end.

It does not seem reasonable, however, to assume that when there was no longer any use for large piping, the purpose of which was to supply steam to these disused machines, that the use of this piping would be continued. Some small reconnections only are necessary to bring about that result, at a total expense of no more than three hundred dollars. Similarly, it was reasonable to assume that in the absence of any operation of machinery during the night hours, arrangements would be made whereby labor would not be maintained to an unnecessary extent, and should it be found more economical to purchase steam during the summer season, that the boiler plant would be then discontinued, and the fire-room labor force dispensed with for that period.

These views rather naturally met with opposition on the part of employees with whose habits and methods they conflicted, but on the whole, the main point determinable by the facts brought out in the trial, is not largely affected by these points, since it was evident from any arrangement of the figures, that the cost of each of the services provided by the plant was excessive, and the general result of the combination of the services, was an unnecessary total expense, which could be reduced by reverting to the method of the purchase of the electrical service, with or without the purchase of steam.

When this point had become so clear as to admit of no further discussion, a new issue was raised by those who represented the interests of the Bureau of Municipal Research. The theory was advanced that the price at which electricity could be purchased, constituted a preferential rate in favor of the City, and that it should not therefore be regarded as a determining figure, and moreover, the rate having come into force since the commencement of the trial, any comparison should be founded not upon that rate, but upon the price which prevailed at the commencement of the trial, which was three cents per kilowatt hour.

This process of reasoning, if it can be so described, coupled with another new assumption, to the effect that the amount of electricity used would be practically the same as it was during the period of the trial (inclusive of the operations of ventilating and lighting the engine room) resulted in an assumption of the cost of purchased electricity exceeding $21,000, which quantity as a matter of actual fact, could then and can now be purchased for less than $14,000. With the more probable lessened demand, the cost would not exceed $12,000. But even with these strained assumptions, and with a total neglect of such conceded elements as the return of taxes paid by the Edison Company to the City, and other elements of cost, the engineering representatives of the Bureau only succeeded in convincing themselves that the purchasable price of electricity under a public service supply should not exceed 1.66 cents per kilowatt hour. As the price at which the energy could be purchased was 1.92

cents per kilowatt hour, the difference they found in favor of the continued operation of the plant was only .26 of a cent per kilowatt hour, amounting on the total consumption they assumed, to only $1,820! Their lame conclusion was thus caught on the horns of a dilemma of their own making.

This result was further determined only upon the bare operating costs, inclusive of no fixed charges, nor any charge for supervision, and irrespective of loss to the city through the change of the rate in the Brooklyn Bridge service. The best that could be said for such a very lamely ascertained result is, that it illustrated the fact that the plant had practically no commercial value, since it was by their own showing, barely competing with the prevailing price of the public supply, and that this result was being effected at an annual expense to the city upon the investment that had been made, amounting to more than $10,000 a year.

The reasonable method of determining the differences in estimated labor was, of course, the operation of another year of trial, during which the facts and figures of expense under another system could have been readily and positively ascertained. This, however, was denied by the action of the majority of the representatives of the parties at issue, those representing the Bureau of Municipal Research, and the City Departments interested, joining in a refusal to recommend the city authorities to authorize the continuance of this comparative trial, in spite of their written agreement to do so. Their motives in thus evading the issue can be left to the judgment of the reader.

The results of this test must therefore stand for themselves, and to any unprejudiced observer who will take the pains to investigate the figures of cost and quantities ascertained in the trial, no other conclusion can possibly be reached than that the expense of operation of this plant, as well as the burdens of its original investment, involve the city in a continuing loss.

The division of the total expense of operating this or any other plant, between such services as it provides, can be so manipulated as to load any one of those services with an undue proportion of the expense, thereby producing an apparent effect of great economy in some other service. But a common-sense disposal of such a method is readily reached by a comparison of the cost of the expensive service with its real commercial value. That value, where a public supply is available, is the price at which equivalent service can be obtained from another source. Thus the steam generated in this building was limited to the value of 42 cents per 1000 pounds weight, at which price it could be purchased in equal quantities from the mains outside this building, and similarly the production of electrical energy could not be of any financial advantage to the city, if its total expense exceeded the price of 1.92 cents per kilowatt hour. Moreover, if the whole combination was more expensive than some other method of operation, the investment and the work of operating the plant were of no commercial or financial value.

The only way in which the effect of these fixed limitations upon the value of the plant could be evaded was by assuming or by actually carrying into effect, some excessive or wasteful use of purchased steam or electricity, or by loading the operation of a reduced number of appliances with some disproportionate extent of labor. But even under such methods all the fixed charges remain as a continual source of expense to the city, and must so continue during the term of existence of the bonds upon the sale of which the money for the investment was obtained.

A very plain understanding of the general results of the investment in this plant is afforded by the summary of the ascertained facts and figures detailed in the preceding chapters. See table page 81. Here we have clearly defined the entire expenses in which the city had been involved in the production of the several services which have been described. Thus the entire cost of the production of the steam can be distributed between the several services in the proportion in which it was used, which is the only way in which it can be fairly divided. If that proportion should result, as it does, in showing that the cost of all the services is excessive, then it would become evident that the whole operation is being conducted at an unnecessary degree of expense.

The only complication that could arise in this simple statement, would be the extent to which the use of steam derived from the engines and pumps can be taken off the cost of their service and charged against the cost of heating the building. But this was disposed of by the facts of the test, and accepted by the Board of Engineers. The steam charged to each service is only that remaining after deducting the exhausted steam used in the work of heating. The cost of each of the several services is relieved of that expense, throwing the expense of the steam upon the work of heating.

The entire quantity of steam used in heating and hot-water work, was barely 40 per cent of the total amount raised during the year, and of this, about 88 per cent was secured in the form of exhausted steam. The allotment of this amount to the credit of each of the several services, was determined by the trial, and finally left a total of 40.5 per cent of all the steam chargeable to the work of heating. This, therefore, carried with it 40.5 per cent of the total expense of producing steam. Therefore the rest of the cost of producing steam, or 59½ per cent, was left to be divided among the other services, of which the test decided the relative proportion of the total steam to be charged to each as follows:

	Per Cent
To electrical generating engines	39.82
To elevator pump operation	12.05
To house water pumping	3.8
To the drinking-water service	2.85
To pumping sewage	0.91

The sub-division of the various costs, therefore, becomes a simple process of division upon these proportions, so far as the steam expense is concerned.

The same common sense methods must apply to the distribution of labor, the expense of which must be divided between the services produced, for they can be disposed of in no other manner. Part of the men's services were so charged by the Department accounts. The remainder must be spread over the services in the same proportion. The other expenses of supplies and repairs were allotted to the several services during the trial.

The total expense was thus divisible in the following fair proportions:

Heating	per cent of total costs	26.9
Electricity	" " " " "	47.0
Elevator	" " " " "	16.45
Water	" " " " "	3.18
Refrigeration	" " " " "	4.5
Sewage	" " " " "	1.97

The alternative system of operation of the several services provided by this plant, which it was agreed by the parties in interest should receive consideration, were never studied in detail in the manner that has been presented in the foregoing chapters. The possibilities and arrangements of alternative operation were ignored, and discussion ranged chiefly around the question of ceasing the operation of only one element in the plant, namely, the electrical generating engines.

It has already been pointed out that a considerable economy could be effected by that simple course of action, but the larger reduction in the cost of the services could be secured through a more complete change of the system of operation, by the purchase of steam from the public service, and by the use of purchased electricity for the operations of elevator service and pumping work as well as lighting. By such a means, a saving of

$15,000 a year could be effected, even with the continued charges for interest and depreciation upon the abandoned boilers and machinery, and also the burden of additional interest and depreciation upon a sum of money required to change the elevators to a more modern electrical form. The expense of operating would then be as follows:

TOTAL EXPENSE OF OPERATION WITH PURCHASED ELECTRICITY AND STEAM

All the steam for house heating, for warm water and for refrigeration purposes, at 42 cents per thousand pounds	$12,910
The present interest and depreciation charges on the boiler plant, continued	5,676
All the electricity required for lighting, and for general power purposes, about 590,000 kilowatt hours per annum, at 1.92c	11,448
The present fixed charges of interest and depreciation on the electric engines and generators, continued	2,256
Operation of the elevators by electricity, involving the fixed charges on the expenditure of $48,000 for electric elevators, the purchase of energy, and the maintenance of the existing interest and depreciation on the hydraulic plant	8,760
The pumping of water by the electrical pump, including the purchase of energy and the continuance of the fixed charges on the existing steam pumping apparatus	898

The refrigerating apparatus, the steam for which is included in the first item, leaving the supplies and repairs, and the existing fixed charges on the apparatus	990
The operation of a sewage pump by electricity, requiring the purchase of electricity and fixed charges on the pump, and the continuance of the existing fixed charges on the present apparatus	463
In none of the foregoing items is any labor included. The operation of the reduced apparatus, most of which would now be entirely automatic, would involve under municipal operation the provision of a chief engineer and one or two assistants. Allow for three men at municipal rates of $5 and $3 per diem	4,015
Total	$47,406

This is $14,823 less than the full expense of operation as determined by the test.

These facts and figures render it impossible to evade the conclusion that the entire operation of these services is excessively expensive. The city is found to be expending a total of $62,229 a year. For that sum it secures three services of importance and three minor services. They are all excessively costly. Several of them could be effected at greatly lessened expense by the use of purchased electricity, and all could be conducted by the use of purchased steam and electricity at twenty-five per cent reduction in expense.

The quantities of steam and of electricity which were used could be substantially reduced if purchased. The fixed charges cannot now be discontinued and must remain as a continuing loss. But the actual disbursements can be reduced by a reduction of labor and fuel, and the purchase of electric energy. These expenses have largely increased since the test by the rising cost of fuel. The most economical methods and care in operation have failed to effect a financial advantage in the continued operation of the plant. Only a stubborn disregard for facts, or misunderstanding of plain figures, stand in the way of relief of expense to the taxpayers of the city.

The discussion of the results of the test resulted in a prolonged delay, and eventual refusal to conduct the comparative trial which had been agreed upon. The New York Edison Company even went so far as to guarantee to provide services of equal extent, at a cost of not exceeding $35,000 a year, or four thousand dollars less than the actual disbursements upon operating costs alone. This guaranteed to the city a direct saving during the period of comparative trial, but in spite of this guarantee the trial was refused.

The conclusion of the investigation, and the facts and comparisons brought forth irresistibly the determination that the investment of municipal funds in the establishment of such a plant as that in this building, even with the subsequent enlargement of its scope of operations, is not of financial advantage to the city, and that the operating costs of such a plant, even when conducted under desirably economical conditions, cannot compete with the reducing prices at which the service of electricity can be purchased from public sources.

The lesson of the failure of this investment, is that the policy of municipal authorities, as well as that of conservative owners of improved real estate, should evidently be seriously to consider, not merely the present circumstances of the operation of a private plant competing with the cost of a public supply, but also the probable future increase in the costs of private operation, as compared with the equally probable reductions in the cost of public service.

The loss that has occurred upon a large number of private plants, the enormous investments in which have become entirely discounted by the progress of these processes, could have been avoided had this reasonable thoughtfulness been more widely exercised. It must be evident to those who look at the matter from an impartial point of view, that the subject should be regarded as one which is a continuing situation over a long period of years, during which, even if some enhanced expense were incurred in the current expenses in the first part of the period, the situation would probably change to the reverse condition towards the end of the period considered. But in any case it would seem to be the part of wisdom to avoid any unnecessary investment of capital in machinery, the best purposes of which are merely competitive, and the entire value of which becomes discounted so soon as the competition in cost has reached the point of balance.

It is certainly a proper course for an engineer to advise his employer or client upon these phases of the purchase of machinery, and to deal with the subject of the use of central station service without the prejudice and selfish interest only too frequently exhibited.

The conclusion from these studies is, therefore, to be very clearly seen in the reduction of expense which is made apparent in the cost of each service under a more modernized system, and by the use of purchased supply of steam and electricity, rather than by the continued operation of machinery under municipal management. But the same effects would be found in private operations, even though the expense of labor would probably be considerably less than under municipal management.

The process is inevitable, in all such combinations of machinery, whereby a gradual change takes place in the surrounding conditions, which have the effect of reducing the value of the output of the combination, which is at all times competing with other methods of accomplishing the same objects. It is this process which terminates the usefulness of operating machinery in advance of the wearing out of its working parts, and places the limit of time upon the effectiveness of such combinations as form the usual isolated plant. Whatever apparatus may be provided under such conditions, the effects which it secures and the output which it provides should be systematically compared with the commercial value of similar service, or equivalent output provided in some other manner.

The developments of mechanical science, and the rapid modification of the habits and needs of modern tenants require changes which must be expected within a comparatively short space of time, and may occur within a very limited period of time. The value of machinery installed for such purposes, even under the most advantageous and economical prevailing conditions always stands liable to be entirely discounted within a few years by the increase in the cost of the materials which it uses, and of the labor which is required for its care, and at the same time, and following a directly opposite process, the product of the plant is undergoing a reduction of its commercial value by declining prices charged for public service, and by alternative methods under which the same results as are afforded by an isolated plant can be attained at lessened cost.

INDEX

A

B

K

R

INDEX

V

W

www.ingramcontent.com/pod-product-compliance
Lightning Source LLC
LaVergne TN
LVHW010552110826
845149LV00003B/634